Block Diagram Implementation for Control Systems

Block Diagram Implementation for Control Systems

Including MATLAB® Programming, Microsoft® Word and PowerPoint Presentation

Ruba Al-Mulla Hummadi
Baghdad University
Iraq

IEEE Press
WILEY

Published by John Wiley & Sons, Inc., Hoboken, New Jersey.

Library of Congress Cataloging-in-Publication Data

Names: Hummadi, Ruba Al-Mulla author
Title: Block diagram implementation for control systems : including MATLAB® programming, Microsoft® Word and PowerPoint presentation / Ruba Al-Mulla Hummadi.
Description: Hoboken, New Jersey : Wiley-IEEE Press, [2025] | Includes index.
Identifiers: LCCN 2025042298 (print) | LCCN 2025042299 (ebook) | ISBN 9781394376049 cloth | ISBN 9781394376063 adobe pdf | ISBN 9781394376056 epub
Subjects: LCSH: MATLAB | Microsoft Word | Microsoft PowerPoint (Computer file) | Control theory | Charts, diagrams, etc. | Flowgraphs
Classification: LCC QA402.3 .H86 2025 (print) | LCC QA402.3 (ebook)
LC record available at https://lccn.loc.gov/2025042298
LC ebook record available at https://lccn.loc.gov/2025042299

Cover Design: Wiley
Cover Image: © Song_about_summer/Shutterstock

Set in 9.5/12.5pt STIXTwoText by Straive, Chennai, India

Dedicated to

the memory of my father, Mustafa Kamal
my family
the memory of my teacher, Prof. Dr. Ihsan Y. Hussain
my colleague, Prof. Dr. Mohsin Al-Shammari
Sandra Grayson
IEEE-Press, and John Wiley & Sons
my uncle, Hazim
the memory of my aunts and uncle
all interested people in Control
Electrical & Mechanical Engineering departments and students
Control & Computer teachers

and

all readers
with love, respect, gratitude and best wishes for a better future.

Contents

About the Author

Ruba Al-Mulla Hummadi, assistant professor from Sacramento. Originally from Iraq. Holds MSc and BSc in Electrical Engineering. Has over 20 years of academic teaching experience at Baghdad University. Taught Electrical Engineering Fundamentals, Electrical Machines and Logic Circuits. Traveled to the USA at the end of 2016. Worked as a tutor at American River College for 3 years.

Preface

Block diagram is one of the most important topics in Control Systems. I found that it is crucial to prepare a book that deals with this topic specifically for students who struggle when solving block diagram reduction and signal flow graph (SFG) problems and put it under their hands.

This book is designed to cover block diagram reduction and SFG, printing, and presentation, in addition to MATLAB codes and pseudocodes. The book is organized as follows:

Chapter 1 defines a block diagram, describes its transfer function and explains the rules that are used in the reduction process.

Chapter 2 deals with MATLAB programming and describes all the commands that are used in the reduction process.

Chapter 3 is the most important part of this book. It contains 24 examples of block diagram reduction with 24 MATLAB codes and pseudocodes to solve specific problems. Some of these examples show more than one method for solving the same problem. I insist on that to let students know how to use specific data from the same given data to make different codes.

Chapter 4 presents some methods that educate the readers to generate a block diagram in Microsoft Word.

Chapter 5 teaches the readers to make a well-done PowerPoint Presentation for any block diagram that can be presented in a discussion, a seminar or a conference. It also describes more than one method in presenting a block diagram relying on what things readers are focusing on.

Chapter 6 deals with signal flow graph terminology. It teaches the reader how to reduce a signal flow graph and how to convert a block diagram into a signal flow graph through 20 examples.

After completion of this book, readers will be able to solve and print any block diagram and signal flow graph, and then prepare a good presentation.

Ruba

Acknowledgment

Thanks, God, for supporting me and giving me the time to complete this project.

A sincere gratitude goes to *Prof. Dr. Mohsin Al-Shammari* for his assistance and helpful notes.

Many thanks go to *Sacramento Public Library clerks* for their assistance in providing some references.

I would like to thank *Sandra Grayson* for always being there for me during the publishing journey. Her assistance and support made my dream come true.

Many thanks go to the publishing house, *IEEE-Press and John Wiley & Sons*, for giving me the chance to be an author and work with them.

1

Block Diagram Introduction and Reduction Rules

1.1 Introduction

A control system is a group of visual components or devices that control or adjust other parts of the system in a specific controlling way. The control system isn't easy to be evaluated or studied, but it can be if we represent it by a block diagram or a signal flow graph. The block diagram is a visual, graphical, or pictorial representation of a system. It consists of a set of blocks that are connected by arrows. Each block has a transfer function that represents a part or device of the system.

In this chapter, you will know what a block diagram means and how it is built. What is a transfer function and how it is represented and evaluated? This chapter also describes the rules of block diagram reduction and teaches how to use these rules in simplifying any block diagram.

Learning Objectives:

- To learn what a block diagram is.
- To know what a transfer function is.
- To learn the rules of block diagram reduction.
- To learn how to apply these rules.

1.2 Definition of Block Diagram

A block diagram as defined by Panda et al. [1] is a pictorial representation of the function performed by each component and flow of signal. In other words, a block diagram is a group of blocks that represent functions and are connected by arrows to show the signal flow. Figure 1.1 shows a sample of a block diagram.

Any block diagram consists of:

- Blocks.
- Summing points.
- Takeoff points.
- Arrows.

Block Diagram Implementation for Control Systems: Including MATLAB® Programming, Microsoft® Word and PowerPoint Presentation, First Edition. Ruba Al-Mulla Hummadi.

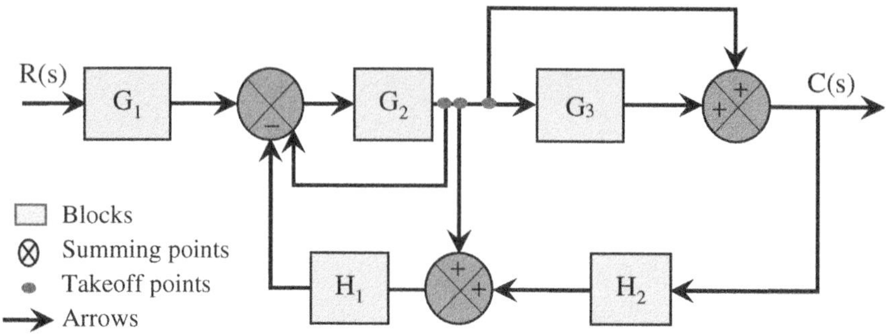

Figure 1.1 Sample of a block diagram.

1.3 Transfer Function

A transfer function could be defined as the mathematical relationship between the input and the output of a system or component, and it is in the Laplace transform [1].

For example, consider the system with the negative feedback in Fig. 1.2.

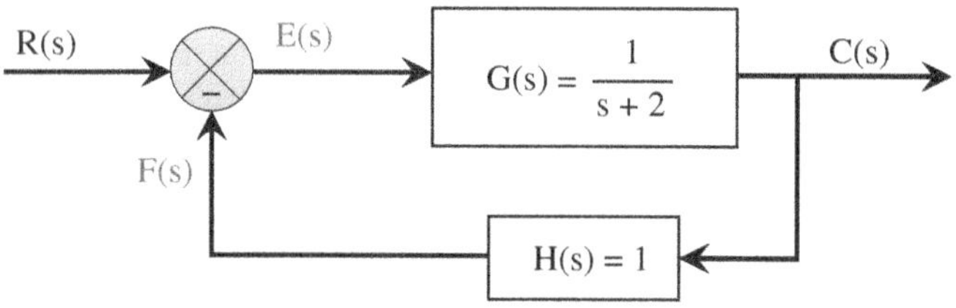

Figure 1.2 System with negative feedback.

It contains two blocks G and H. The input and the output for block G are E(s) and C(s). The input and the output for block H are C(s) and F(s). Also, we have the reference input R(s). All we need to do is to get the overall transfer function for this system. So, we will start from the output C(s) and write an equation for it and then substitute the variables until we get an equation in terms of C(s) and R(s) only.

$$C(s) = E(s)\,G(s) \tag{1.1}$$

$$E(s) = R(s) - F(s)\ \&\ \ F(s) = C(s)\,H(s)$$

$$E(s) = R(s) - C(s)\,H(s) \tag{1.2}$$

By substituting Eq. (1.2) in Eq. (1.1), to get:

$$C(s) = [R(s) - C(s)\,H(s)]\,G(s) \tag{1.3}$$

$$= R(s)\,G(s) - C(s)\,H(s)\,G(s)$$

$$C(s) + C(s)\,H(s)\,G(s) = R(s)\,G(s)$$

$$C(s)[1 + H(s)\,G(s)] = R(s)\,G(s)$$

$$\frac{C(s)}{R(s)} = \frac{G(s)}{1 + H(s)\,G(s)} \tag{1.4}$$

Substitute by G(s) = 1/s + 2 and H(s) = 1, to get:

$$\frac{C(s)}{R(s)} = \frac{\frac{1}{s+2}}{1+1\left(\frac{1}{s+2}\right)} = \frac{1}{s+3} \tag{1.5}$$

1.4 Rules of Block Diagram Reduction

Many control books contain the rules of reduction either included in the explanation or separated in an appendix at the end of the book. Here, what we want to do is to put these rules under the reach of any student and let him/her put these rules near him/her during solving any block diagram reduction problem. Table 1.1 shows the reduction rules with their block representations and explanations. The figures for rules 1–9 are taken from Panda et al. [1], Jairth [2], and Ogata [3] and the rest are drawn to match the style of this book. The results after applying the rules are also shown. Go to the mentioned reference [3], Ghosh [4] for more details about how each rule was driven.

Table 1.1 Block diagram reduction rules.

	Description	Original and equivalent block diagrams	Explanations
Rule 1	Blocks in series or cascade	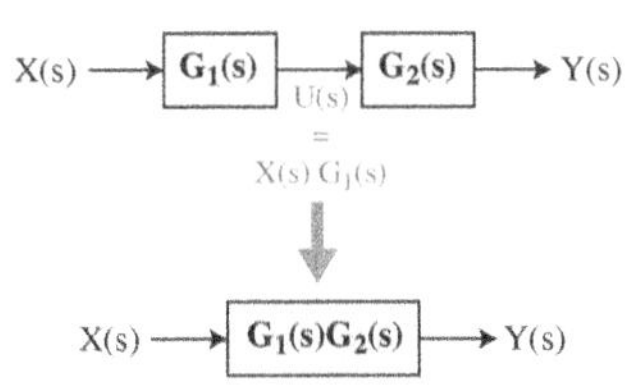	The overall transfer function is the product of transfer functions for all individual blocks that are connected in series
Rule 2	Blocks in parallel or eliminating a forward loop	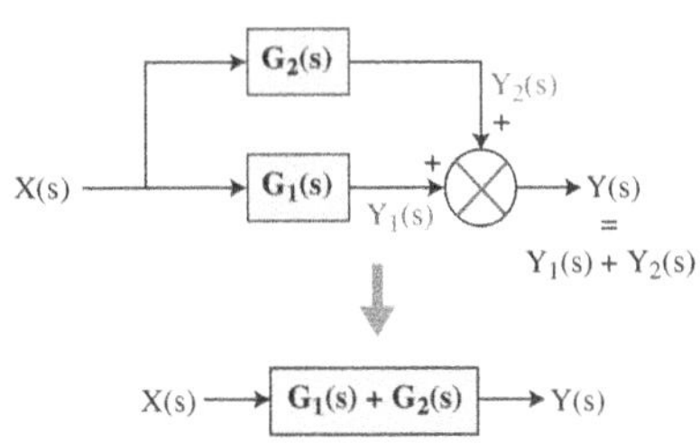	The overall transfer function is the summing or differencing of transfer functions for all individual blocks that are connected in parallel depending on the sign + or −
Rule 3	Moving a takeoff point after a block	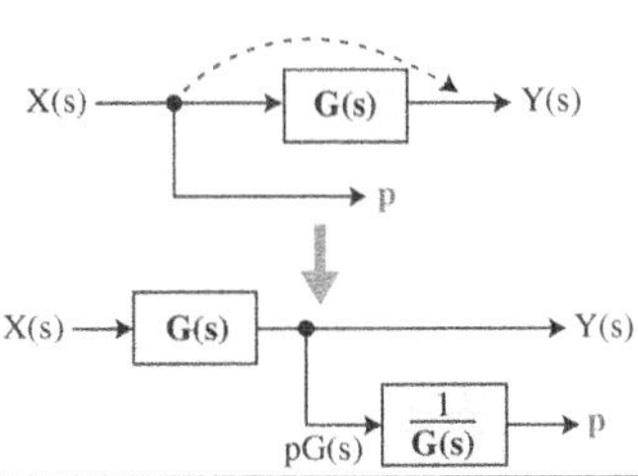	Divide on the transfer function of the block

(Continued)

Table 1.1 (Continued)

	Description	Original and equivalent block diagrams	Explanations
Rule 4	Moving a takeoff point before a block	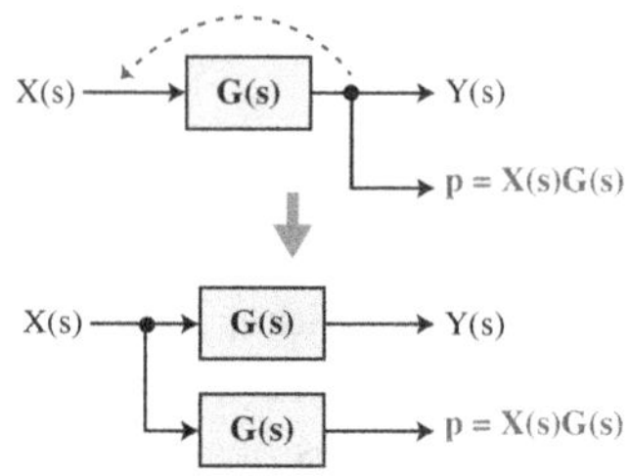	Multiply by the transfer function of the block
Rule 5	Moving or shifting a summing point or an adder after a block	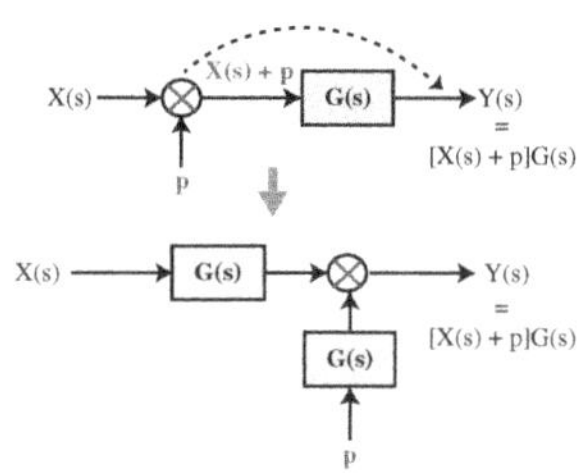	Multiply by the transfer function of the block
Rule 6	Moving or shifting a summing point or an adder before a block	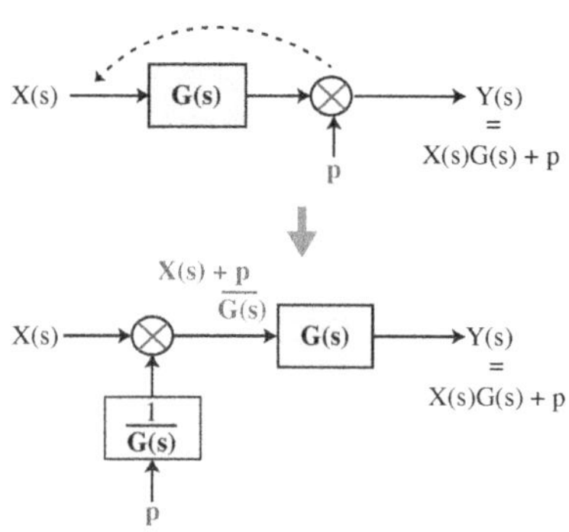	Divide on the transfer function of the block
Rule 7	Interchange of the summing point or rearrangement of adders	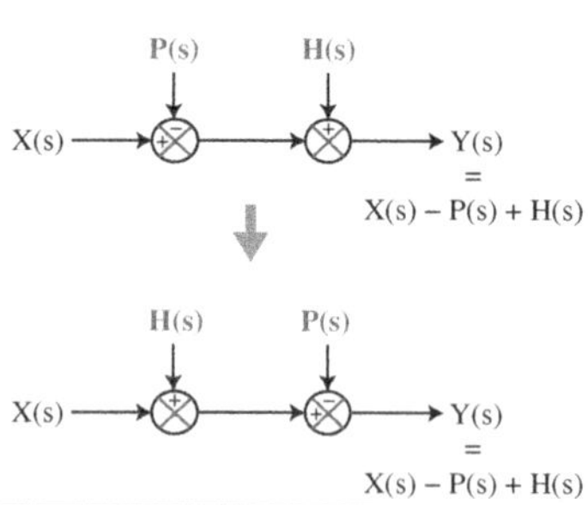	Just change the places of adders

Table 1.1 (Continued)

	Description	Original and equivalent block diagrams	Explanations
Rule 8	Splitting/combining the summing point	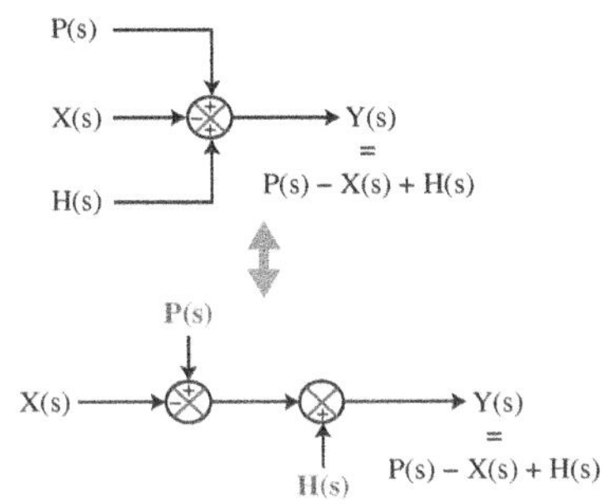	
Rule 9	Eliminating the feedback loop	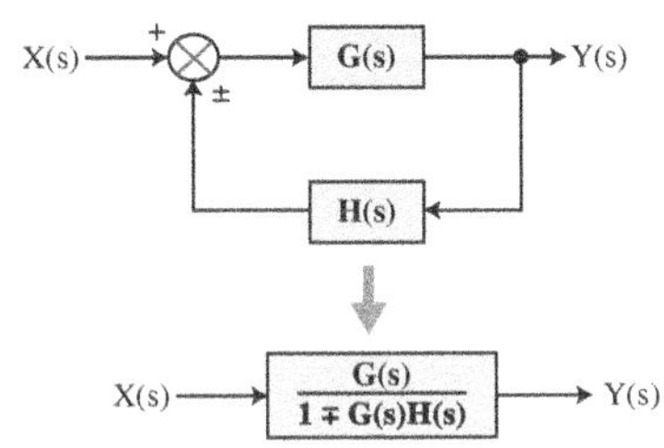	
Rule 10	Removing a block from a forward path	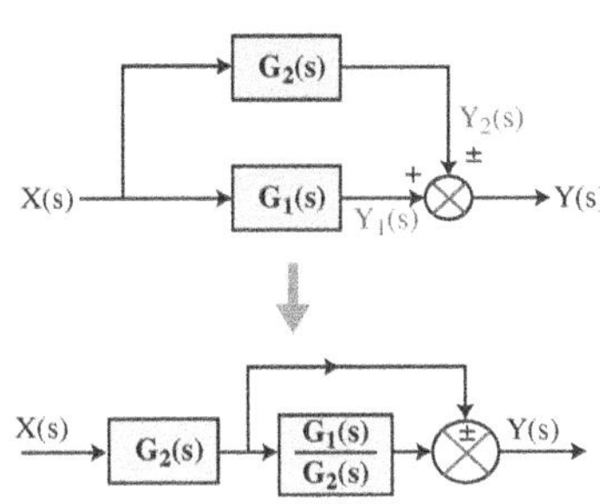 $\frac{X}{Y} = G_2\left(\frac{G_1}{G_2} \pm 1\right)$	
Rule 11	Removing a block from a feedback loop	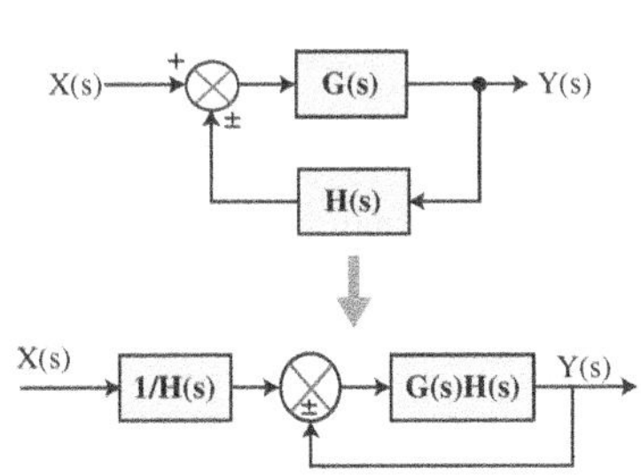 $\frac{X}{Y} = \frac{1}{H}\left(\frac{GH}{1 \mp GH}\right)$	

Review Questions

1. Define a block diagram.
2. What does a block diagram consist of?
3. If G_1 and G_2 are cascade blocks, what is the equivalent block?
4. What do you do if you moved a takeoff point before a block?
5. How can you remove a block from a forward path?
6. How can you eliminate feedback?
7. How do you rearrange two adders?
8. How do you split or combine two summing points?

2

MATLAB®

2.1 Introduction

MATLAB® is software from MATHWORKS. It offers a platform used to program and analyze engineering designs. It is used for many purposes such as analyzing data, developing algorithms, and creating models and applications.

MATLAB is the abbreviation of Matrix Laboratory. This means that every entered data into the MATLAB environment must be in a matrix form. To start working on MATLAB, the user should create a new M-file and use either inbuilt functions or create ones. Inbuilt functions are functions that are already defined in MATLAB such as `series, parallel, and Cloop`. Other functions that are created by the user are user-defined functions. For more information on MATLAB, read Palm III [5] and Gilat [6] or, watch a video on YouTube.

This chapter focuses on how to reduce a block diagram in MATLAB. How to create a new M-file and add more work on an existing file. What commands should be used in the reduction process and how.

Learning Objectives:

- To learn how to work on MATLAB.
- To learn how to apply the appropriate MATLAB command on each block diagram reduction stage or step.
- To learn how to write a MATLAB script for a block diagram reduction step.
- To learn how to write a pseudocode.
- To learn how to convert a pseudocode into a MATLAB code.

2.2 To Start with MATLAB®

- Install MATLAB® software on your computer and get the MATLAB icon on your desktop.

Block Diagram Implementation for Control Systems: Including MATLAB® Programming, Microsoft® Word and PowerPoint Presentation, First Edition. Ruba Al-Mulla Hummadi.

- Double-click on this icon to enter the MATLAB® program as shown in Fig. 2.1. The MATLAB window is divided into three parts in default format: the command window, the workspace area, and the current folder area.
  - The command window is the most important area because it is a multifunctional area. First, it can be used to execute any script or any MATLAB command. Next, it can be used to ask for MATLAB help. Finally, it can be used as a scientific calculator.
  - The workspace area displays the data and the results.
  - The current folder area shows the position of the M- file on a PC.

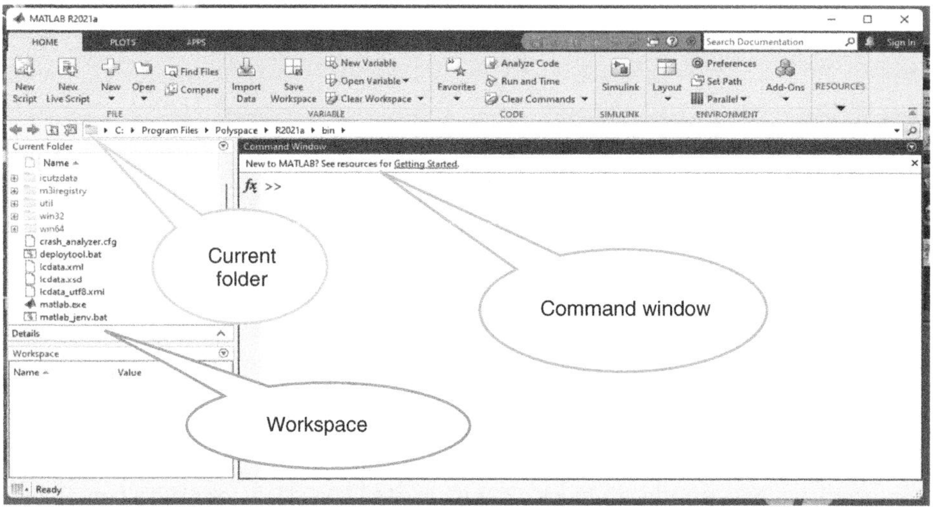

Figure 2.1 The MATLAB screen.

- The editor window is the window where the programmer can write his script. It is highly recommended to open a new window for this editor so that the programmer can have a large area to work on as shown in Fig. 2.2. For more info, see Chapter 1 [6].

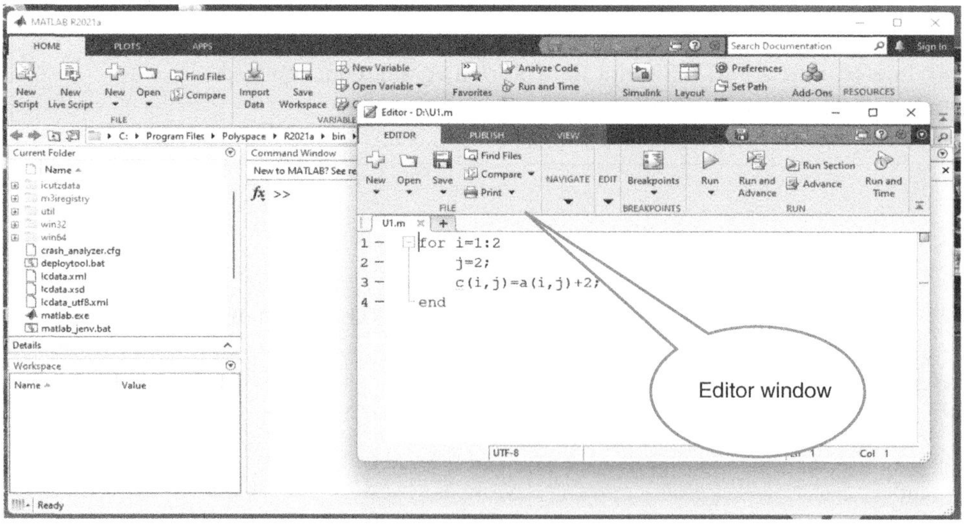

Figure 2.2 The MATLAB screen with the editor window.

2.3 M-File Definition

M-file(s) is/are the program(s) that be created by the user(s) or the programmer(s) and written in the editor window.

2.4 Create a New M-File (New Script)

To create or open a new M-file or script; open MATLAB and click on the HOME tab, then on the FILE group go to New and open its menu, then select and click Script as shown in Figs. 2.3 and 2.4.

Another Way to Open a New M-file

Go to FILE group, and click New Script, then an Editor Window will pop up on the screen as shown in Fig. 2.5.

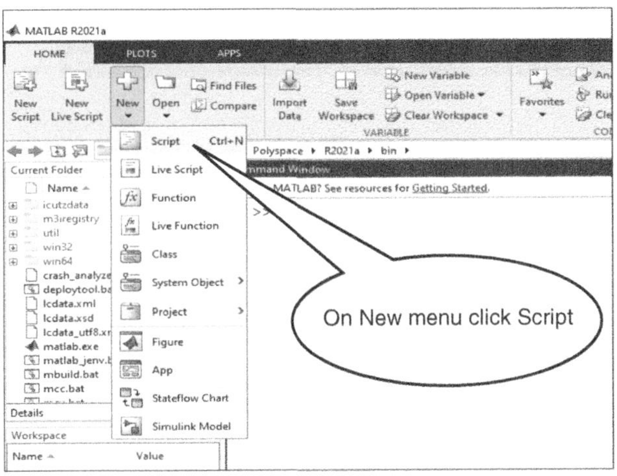

Figure 2.3 Click Script from the New menu.

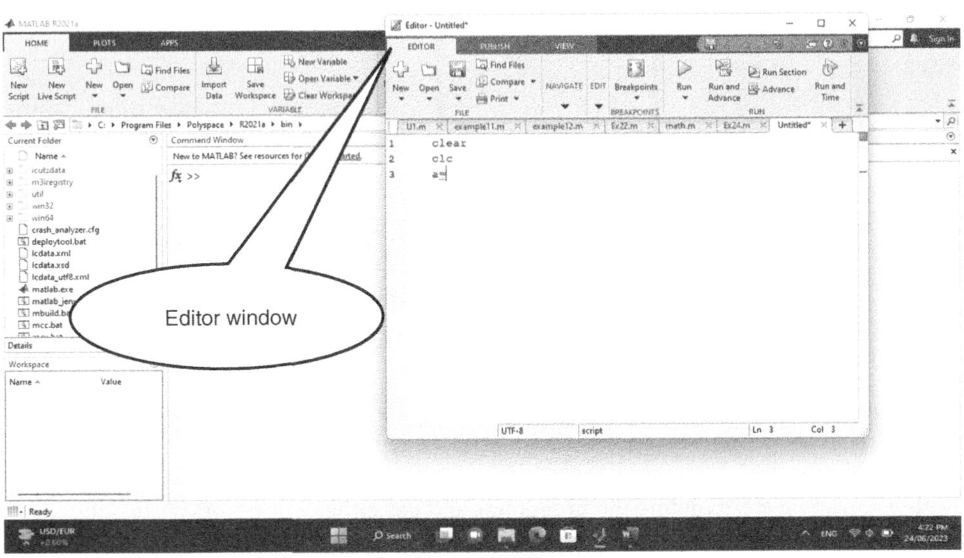

Figure 2.4 Editor window.

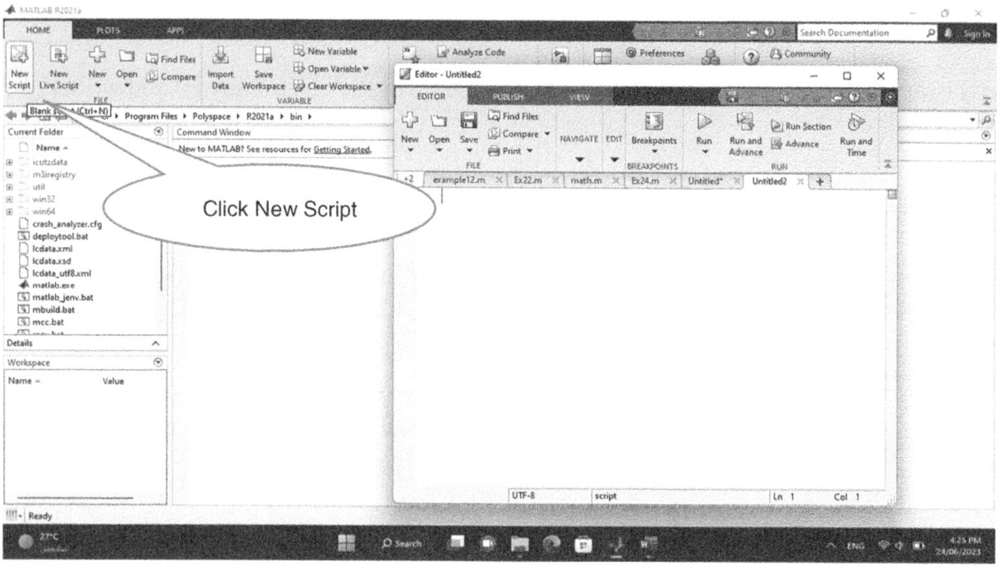

Figure 2.5 New Script and editor window.

2.5 Write a MATLAB® Script

After opening the Editor window, follow the three-step process of writing a MATLAB script as shown in Fig. 2.6.

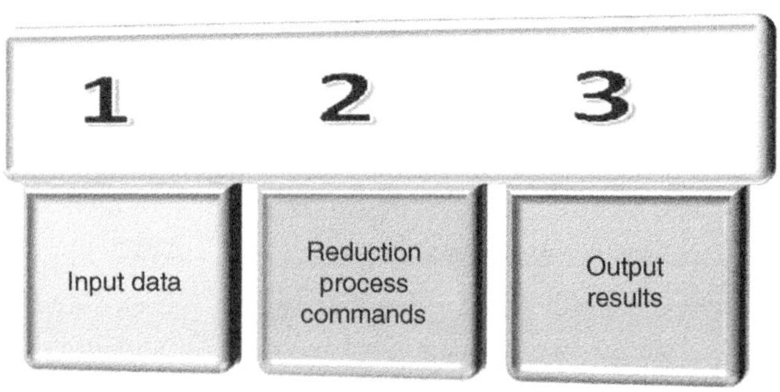

Figure 2.6 How to write a MATLAB script file.

Write "`clear`" and "`clc`" before starting any program to delete any previous execution and clean the screen.

1- **Input data:**
   Input all G's and H's block transfer functions. For example, $G = 1/(s+1)$, use either way to enter the transfer function.
   - Defining numerator and denominator:
     Use either "`num`" or "`n`" to define the numerator and use either "`den`" or "`d`" to define the denominator of each block.

```
numG=[1]; denG=[1   1];
or
nG=[1]; dG=[1   1];
```

- Use `tf` command:

```
s=tf('s');
G=1/(s+1);
```

2- **Reduction process commands:**
   - Reduce or simplify the block diagram stage-by-stage using the appropriate MATLAB commands.
   - Use Table 2.1 to apply the appropriate MATLAB command for the reduction process.

Table 2.1 MATLAB commands that used in the reduction process.

1-	Command	`Series`
	Figure	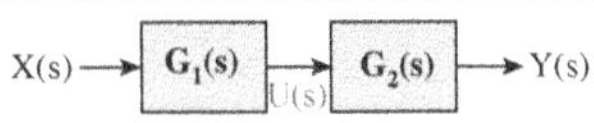
	Rule	`[n,d]=series(nG1,dG1,nG2,dG2)`
	Description	[output numerator, output denominator] =MATLAB command (numerator of G_1, denominator of G_1, numerator of G_2, denominator of G_2)
		If G & H are transfer functions use `GT=series(G1,G2)`
2-	Command	`Parallel`
	Figure	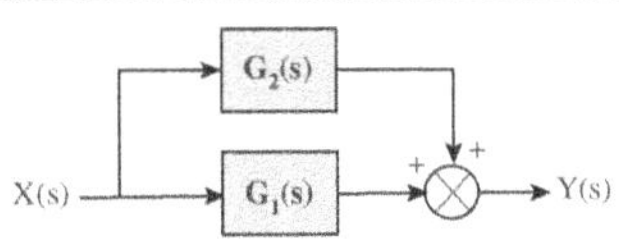
	Rule	`[n,d]=parallel(nG1,dG1,nG2,dG2)`
	Description	[output numerator, output denominator] = MATLAB command (numerator of G_1, denominator of G_1, numerator of G_2, denominator of G_2)
		If G & H are transfer functions use `GT=parallel(G1,G2)`
3-	Command	`feedback`
	Figure	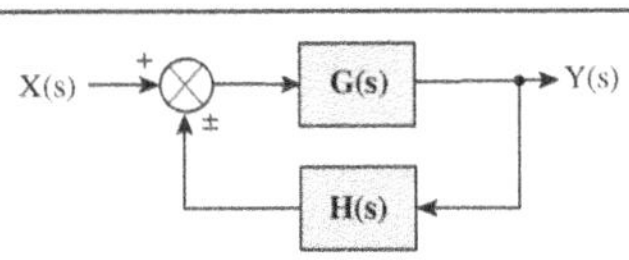
	Rule	`[n,d]=feedback(nG,dG,nH,dH, ±1)`
	Description	[output numerator, output denominator] = MATLAB command (numerator of G, denominator of G, numerator of the feedback H, denominator of H, feedback sign) The negative feedback is the MATLAB default. Use `[n,d]=feedback(nG,dG,nH,dH)` or `[n,d]=feedback(nG,dG,nH,dH,-1)` for negative feedback Use `[n,d]=feedback(nG,dG,nH,dH,+1)` for positive feedback
		If G & H are transfer functions use `G1=feedback(G,H)` for negative feedback `G1=feedback(G,H,+1)` for positive feedback

(Continued)

Table 2.1 (Continued)

4-	Command	`cloop` Cloop means closed loop. In other words, there is no block in the feedback loop (H = 1).
	Figure	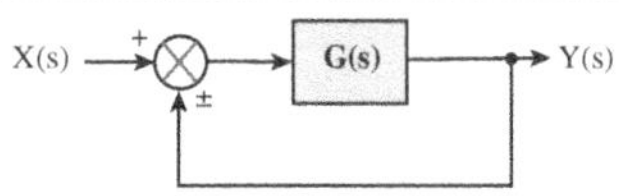
	Rule	`[n,d]=cloop(nG,dG,±1)`
	Description	[output numerator, output denominator] = MATLAB command (numerator of G, denominator of G, close loop sign)
		If G & H are transfer functions, use `GT=cloop(G,±1)`
5-	Command	`Printsys`
	Figure	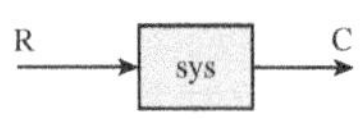
	Rule	`printsys(n,d)`
	Description	MATLAB command (numerator of the system transfer function, denominator of the system transfer function)
6-	Command	`conv`
	Figure	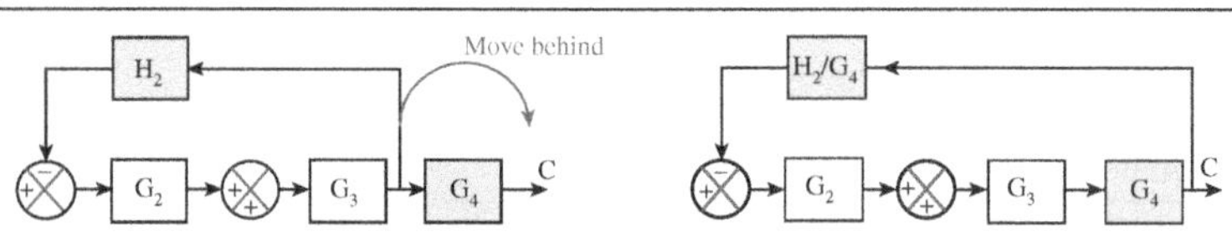
	Rule	`n=conv(nH2,dG4), d=conv(dH2,nG4)` `printsys(n,d)`
	Description	When you have two transfer functions divided on each other like Eq. (2.1), use the `conv` command to get the result as numerator and denominator. Then use `printsys` command to write the result as a transfer function. $$\frac{H_2}{G_4} = \frac{\frac{nH_2}{dH_2}}{\frac{nG_4}{dG_4}} = \frac{nH_2}{dH_2} * \frac{dG_4}{nG_4} \quad (2.1)$$
7-	Command	`disp`
	Rule	`disp('output transfer function')`
	Description	MATLAB command ('text')

3- **Output results:**

- To display the results, follow either way:
  - Don't put a semicolon (;) at the end of your script file line.
  - Use "`printsys`" command to display what you want.

After writing the program, save it under a chosen name, then execute it. To detect errors easily, it is highly recommended to execute the program after each step. The entered Data into MATLAB should be in matrix form. Matrix is explained in Section 2.6.

MATLAB can assist you when you need help understanding a command or when you don't know how to use a command. Write help followed by the command name in the Command Window, then press enter. A full explanation about the command will appear on the screen as shown in Fig. 2.7.

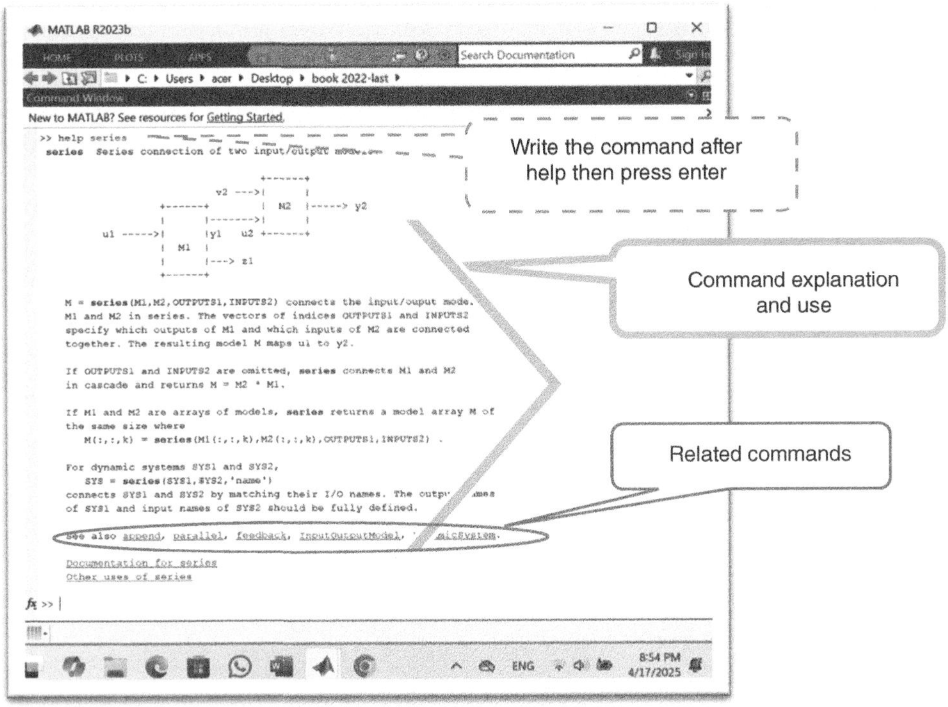

Figure 2.7 The use of MATLAB help.

2.6 Matrix

Matrix is a group of elements that are organized in rows and columns. Those elements may be numbers, symbols, equations, or mix. Table 2.2 shows examples of matrices that were written in MATH and their MATLAB representation.

Table 2.2 Samples of matrices and their MATLAB representation.

	MATH	MATLAB
1-	$a = [1]$	`a=[1]`
2-	$B = [1\ \ 2]$	`B=[1 2]`
3-	$c = \begin{bmatrix} 1 & 2 \\ 3 & 4 \end{bmatrix}$	`c=[1 2;3 4]`
4-	$d = \begin{bmatrix} 0 & \sin(\pi/2) & X^2 \\ 1+2X & 5 & 6\pi \\ 1 & 3 & -4.2 \end{bmatrix}$	`d=[0 sin(pi/2) X^2; (1+2*X) 5 6*pi; 1 3 -4.2]`
5-	$e = \begin{bmatrix} 1 & -3 & 5x \\ 4x^2 & 8 & 0 \end{bmatrix}$	`e=[1 -3 5*x;4*x^2 8 0]`

2.7 Saving an M-File

To save a new M-file, in the Editor window under EDITOR, go to the FILE group, go to Save and open its menu, then click Save as if this is the first time you open a file, as shown in Fig. 2.8, name your file and save it on your computer. Note to click Save after adding more work to any old file.

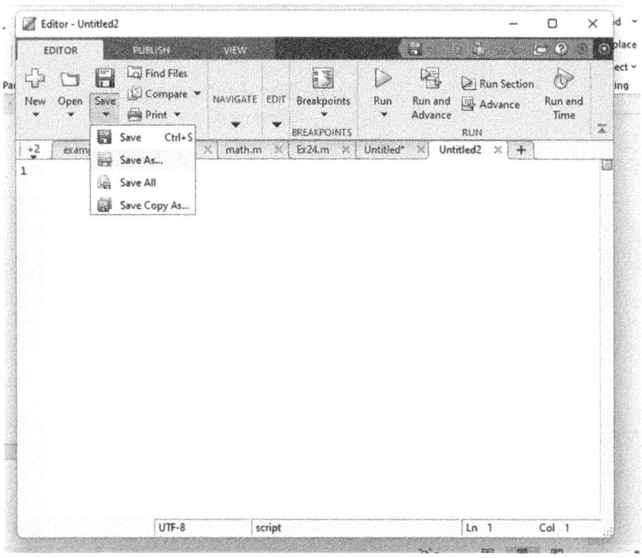

Figure 2.8 Save as ... in the Editor window.

2.8 Executing or Running an M-File

On the editor window and under the EDITOR tab, click the Run icon to execute an M-file as shown in Fig. 2.9.

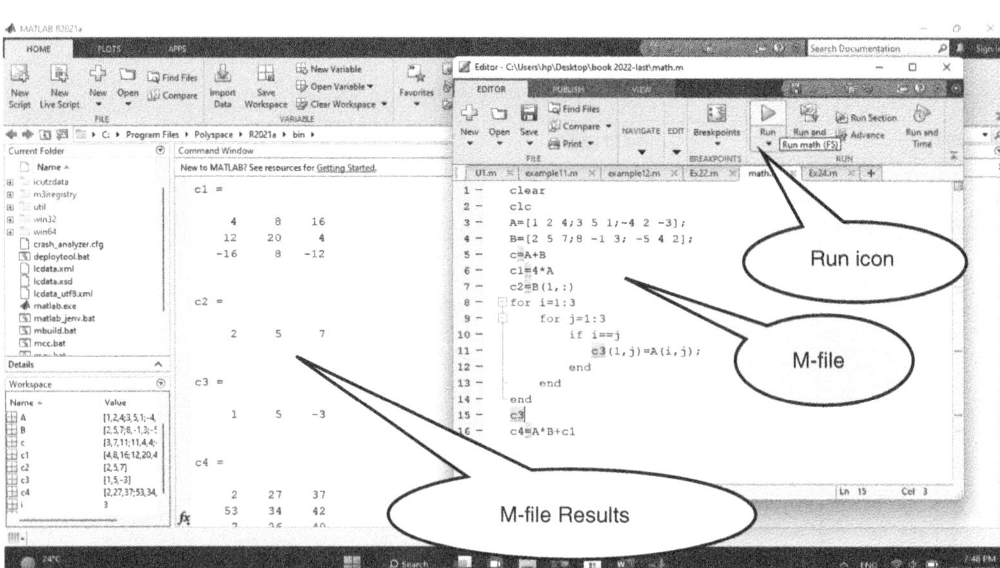

Figure 2.9 Run an M-file.

2.9 How Can MATLAB® Detect Errors and How It Can't

When the programmer executes his program, MATLAB shows a red error message in the command window. The programmer should read this massage and correct his program, then execute it one more time. MATLAB will continue showing errors until there is no error anymore as shown in the following example.

Example 2.1

$$\text{Given } A = \begin{bmatrix} 1 & 2 & 4 \\ 3 & 5 & 1 \\ -4 & 2 & -3 \end{bmatrix} \text{ and } B = \begin{bmatrix} 2 & 5 & 7 \\ 8 & -1 & 3 \\ -5 & 4 & 2 \end{bmatrix},$$

find the following:

1- A+B,
2- 4*A,
3- 1st row of B,
4- The diagonal of A,
5- The results of A*B+4*A.

Solution:

Suppose the MATLAB code was written like this:

```
clear
clc
% % file name: math.m
A=[1 2 4;3 5 1;-4 2 -3];
B=[2 5 7;8 -1 3; -5 4 2];
c=A+b
c1=4A
c2=B(:,1)
for i=1:3
    for j=1:3
        if i=j
           c3(1,j)=A(i,j);
        end
    end
end
c3
c4=A*B+C1
```

The program has 4 errors

For more info about for–end & if–end structure see the appendix

MATLAB will display a red error message after each time of execution. MATLAB will give results when there is no error anymore as it is illustrated in Table 2.3.

Table 2.3 The MATLAB error messages.

Run time	MATLAB error massage	Explanation	Correction
1- First time run	`Unrecognized function or variable 'b'.`	Write B instead of b because we don't have b in this example.	`B`
2- Second time run	`File: math.m Line: 7 Column: 5 Invalid expression. Check for missing multiplication operator, missing or unbalanced delimiters, or other syntax error. To construct matrices, use brackets instead of parentheses.`	MATLAB tells about the position (line & column) of the error; use the multiplication operator * and write 4*A instead of just 4A.	`4*A`
3- Third time run	`File: math.m Line: 11 Column: 13 Incorrect use of '=' operator. Assign a value to a variable using '=' and compare values for equality using '=='.`	There is an error in the MATLAB rules; use == instead of =.	==
4- Fourth time run	`No massage`	Get the results as shown below.	

After Execution, The Results Will Be:

```
c =

     3     7    11
    11     4     4
    -9     6    -1


c1 =

     4     8    16
    12    20     4
   -16     8   -12


c2 =

     2
     8
    -5


c3 =

     1     5    -3


c4 =

     2    27    37
    53    34    42
     7   -26   -40
```

After checking MATLAB results with the programmer's results, the programmer will see that c2 is not like the one that he acquired by hand.

In this case, MATLAB gave the results and didn't show an error message. The reason is the error wasn't in **MATLAB rules**; it was the programmer's error. In other words, it was an error in the programming way. Table 2.4 shows the difference between the two MATLAB expressions with results and clarifications.

Table 2.4 The difference between two MATLAB expressions.

Expression	Results	Meaning	Clarifications
`c2=B(:,1)`	c2 = 2 8 -5	: means all rows of B, 1 means the first column of B	This expression shows a column of B
`c2=B(1,:)`	c2 = 2 5 7	1 means the first row of B, : means all the columns of B	This expression shows a row of B

So, after correcting the expression for c2, the MATLAB code for this example will be as shown below:

```
clear
clc
% % file name: math.m
A=[1 2 4;3 5 1;-4 2 -3];
B=[2 5 7;8 -1 3; -5 4 2];
c=A+B
c1=4*A
c2=B(1,:)
for i=1:3
    for j=1:3
        if i==j
           c3(1,j)=A(i,j);
        end
    end
end
c3
c4=A*B+c1
```

Correct code

And The Correct Solution Will Be:

```
c =

     3     7    11
    11     4     4
    -9     6    -1
```

```
c1 =

     4     8    16
    12    20     4
   -16     8   -12

c2 =

     2     5     7

c3 =

     1     5    -3
c4 =

     2    27    37
    53    34    42
     7   -26   -40
```

Example 2.2 Given $A = \begin{bmatrix} 2 & 4 & -10 \\ 3 & -5^2 & 1 \end{bmatrix}$, $B = \begin{bmatrix} 3 & 7 & 9 \\ 1 & -2 & -5 \end{bmatrix}$ & $D = \begin{bmatrix} -8 \\ \sin\left(\frac{\pi}{2}\right) \\ 3 \\ 6 \end{bmatrix}$

1- To create matrix C (4*3), place matrix A in the first and second rows, followed by matrix B in the third and fourth rows.
2- To create matrix E (4*4), use matrix D as the fourth column of matrix C.
3- Determine the determinant of matrix E.
4- Find the inverse of matrix E.

Solution:

The given matrices are formed as:

$$A = \begin{bmatrix} A_{11} & A_{12} & A_{13} \\ A_{21} & A_{22} & A_{23} \end{bmatrix}, B = \begin{bmatrix} B_{11} & B_{12} & B_{13} \\ B_{21} & B_{22} & B_{23} \end{bmatrix} \& D = \begin{bmatrix} D_{11} \\ D_{21} \\ D_{31} \\ D_{41} \end{bmatrix}$$

Then the manual solution for the matrices C & E will be:

$$C_{4*3} = \begin{bmatrix} C_{11} & C_{12} & C_{13} \\ C_{21} & C_{22} & C_{23} \\ C_{31} & C_{32} & C_{33} \\ C_{41} & C_{42} & C_{43} \end{bmatrix} = \begin{bmatrix} A_{11} & A_{12} & A_{13} \\ A_{21} & A_{22} & A_{23} \\ B_{11} & B_{12} & B_{13} \\ B_{21} & B_{22} & B_{23} \end{bmatrix},$$

$$E_{4*4} = \begin{bmatrix} E_{11} & E_{12} & E_{13} & E_{14} \\ E_{21} & E_{22} & E_{23} & E_{24} \\ E_{31} & E_{32} & E_{33} & E_{34} \\ E_{41} & E_{42} & E_{43} & E_{44} \end{bmatrix} = \begin{bmatrix} C_{11} & C_{12} & C_{13} & D_{11} \\ C_{21} & C_{22} & C_{23} & D_{21} \\ C_{31} & C_{32} & C_{33} & D_{31} \\ C_{41} & C_{42} & C_{43} & D_{41} \end{bmatrix}$$

The MATLAB code for this example will be:

```
clear
clc
% input data
A=[2 4 -10;3 -5^2  1];
B=[3 7 9;1 -2 -5];
D=[-8; sin(pi/2); 3; 6];
% create Matrix C from A & B
for i=1:2
    for j=1:3
        C(i,j)=A(i,j);
        C(i+2,j+2)=B(i,j);
    end
end
C
% Create Matrix E by adding another column (D) to C
for i=1:4
    for j=1:4
        if j==1
           E(i,j)=D(i,j);
        end
        E(i,j)=C(i,j);
    end
end
E
```

After Program Execution:

```
C =
     2     4   -10     0     0
     3   -25     1     0     0
     0     0     3     7     9
     0     0     1    -2    -5

E =
     2     4   -10     0
     3   -25     1     0
     0     0     3     7
     0     0     1    -2

q =
  806.0000
```

```
q1 =
     0.4032    0.0645    0.6104    2.1365
     0.0484   -0.0323    0.0794    0.2779
          0         0    0.1538    0.5385
          0         0    0.0769   -0.2308
```

When contrasting the outcomes of Matlab's analysis with those achieved by hand, we discover that they are not match for some reasons.

- The index j+2 in the expression `C(i+2,j+2)=B(i,j)` is incorrect because when j=1, the programmer intended to create elements from C to match those in B but ended up accessing the third column instead of remaining in the first one. This mistake is classified as a user error. There is nothing wrong with MATLAB's rules. Therefore, the correct expression should be `C(i+2,j)=B(i,j)` as demonstrated in the following table.

Wrong code	Right code

`for i=1:2`
`    for j=1:3`
`        C(i,j)=A(i,j);`
`        C(i+2,j+2)=B(i,j);`
`    end`
`end`
`C`
	`for i=1:2`
	`    for j=1:3`
	`        C(i,j)=A(i,j);`
	`        C(i+2,j)=B(i,j);`
	`    end`
	`end`
	`C`

- The second part of the MATLAB code is also inaccurate because the expression `E(i,j)=D(i,j)` is a conditional statement that only executes when j is equal to 1, while the expression `E(i,j)=C(i,j)` executes every time. Therefore, when we assign E_{11} to be equal to D_{11}, the program will subsequently overwrite it with C_{11}, making it all incorrect. We need to adopt a different programming approach by using the expression `E(i,j)=C(i,j)` for the first three columns, and the fourth column should be set to the D matrix, as illustrated in the table below.

Wrong code	Right code

`for i=1:4`
`    for j=1:4`
`        if j==1`
`            E(i,j)=D(i,j);`
`        end`
`        E(i,j)=C(i,j);`
`    end`
`end`
`E`
	`for i=1:4`
	`    for j=1:3`
	`        E(i,j)=C(i,j);`
	`    end`
	`    E(i,4)=D(i,1);`
	`end`
	`E`

The corrected code wil be:

```
clear
clc
% input data
A=[2 4 -10;3 -5^2  1];
B=[3 7 9;1 -2 -5];
D=[-8; sin(pi/2); 3; 6];
% create Matrix C from A & B
for i=1:2
    for j=1:3
        C(i,j)=A(i,j);
        C(i+2,j)=B(i,j);
    end
end
C
% Create Matrix E by adding another column (D) to C
for i=1:4
    for j=1:3
        E(i,j)=C(i,j);
    end
    E(i,4)=D(i,1);
end
E
q=det(E)
q1=inv(E)
```

After Execution The Results Will Be:

```
C =
     2     4   -10
     3   -25     1
     3     7     9
     1    -2    -5

E =
     2     4   -10    -8
     3   -25     1     1
     3     7     9     3
     1    -2    -5     6

q =
      -14624
```

```
q1 =
     0.1224     0.0610     0.1677     0.0692
     0.0115    -0.0328     0.0219     0.0098
    -0.0346     0.0096     0.0353    -0.0654
    -0.0454    -0.0131     0.0088     0.1039
```

2.10 Pseudocode Definition

Pseudocode is the steps that guide writing a program in any required computer language.

2.11 Write a Pseudocode

There is no specific language to write pseudocode. The programmer can use his language. As shown in the following example.

2.12 Converting a Pseudocode into MATLAB Code

It is so easy to convert a pseudocode into MATLAB code. The programmer just needs to see what is the step in a pseudocode and write it in the MATLAB language as shown in Example 2.3.

Example 2.3 Write a pseudocode for the block diagram of Fig. 2.10 and then convert it into a MATLAB code.

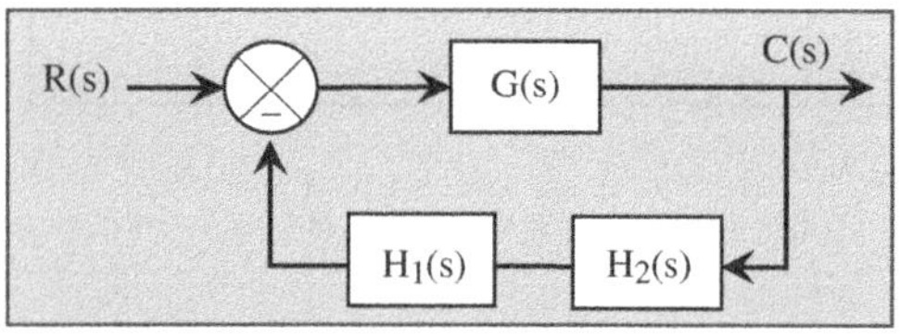

$$G_1 = \frac{1}{s+1},\quad H_1 = \frac{1}{s+2},\quad H_2 = 2$$

Figure 2.10 Block diagram of Example 2.3.

Solution:

To write a pseudocode, put the steps that guide the solution of the block diagram. In Fig. 2.10, H_1 & H_2 blocks are connected in series. The elimination of the feedback loop formed by H_1H_2 gives the overall system transfer function. So, when the programmer writes these steps, they will easily guide him to write the MATLAB code as shown in Table 2.5.

Table 2.5 The pseudocode and its conversion to MATLAB code.

Pseudocode	MATLAB code
1- Remove and clean any pervious program execution.	`clear`
2- Clean your screen.	`clc`
3- Name your file.	`% file name: EX2_3.m`

4- Define numerators and denominators for all G and H's blocks.	`nG=[1];dG=[1 1];`
	`nH1=[1];dH1=[1 2];`
	`nH2=[2];dH2=[1];`

5- Series (H_1 & H_2)	`[n,d]=series(nH1,dH1,nH2,dH2)`
6- Feedback (G, H_1H_2)	`[n1,d1]=feedback(nG,dG,n,d,-1)`
7- Display C(s)/R(s)	`Printsys(n1,d1)`

Review Questions

1 What is the purpose of the Command Window?

2 Where does the user write his script?

3 Enter matrix $A = \begin{bmatrix} \frac{1}{s+2} & 2 \\ \pi & s+3 \\ 1 & -4 \end{bmatrix}$ into MATLAB.

4 How do you use the feedback command?

5 How do you use the series command?

6 What is the difference between feedback and Cloop?

7 What happens if you didn't put a semicolon at the end of your M-file line?

8 What is a pseudocode?

9 How to convert pseudocode into MATLAB code?

10 Convert this pseudocode step **feedback (G,H,−1)** into MATLAB code knowing that $G = \frac{1}{s+10}$, $H = s + 2$.

3

Examples of Block Diagram Reduction

3.1 Introduction

This chapter is the heart of this book; it contains around 24 examples of block diagram reduction. These examples are taken from the mentioned references and solved in different ways. There is a start mark and end mark for each way of solution to let the reader understand the starting and the completion of the mentioned method. MATLAB codes, pseudocodes, and transfer functions were shown after each stage of reduction to confirm understanding in some examples. In others, the reader may see the complete MATLAB code at the end of the example. The book also shows the solution after each MATLAB code. Some activities need to be solved at the end of some examples. Also, there are some more practice parts. These are solved examples in the mentioned references and are related to the examples in this book. So, the reader can go through these examples for more understanding.

Learning Objectives:

- To learn how to reduce a block diagram.
- To learn how to apply block diagram reduction rules to each case.
- To learn how to write a pseudocode.
- To learn how to write MATLAB code.
- To learn how to convert pseudocode into MATLAB code.

3.2 Examples of Block Diagram Reduction

In this section, the reader will see 24 examples solved in different ways. Focus on how the rules of reduction are applied in each case and how the block diagram is reduced. Always return to Chapter 1 to review these rules and see how they are applied in each case.

Block Diagram Implementation for Control Systems: Including MATLAB® Programming, Microsoft® Word and PowerPoint Presentation, First Edition. Ruba Al-Mulla Hummadi.

Example 3.1 Reduce the block diagram of Fig. 3.1, then obtain the overall system transfer function [2, p: 110, problem 5.1]; [4, p: 212, Q7].

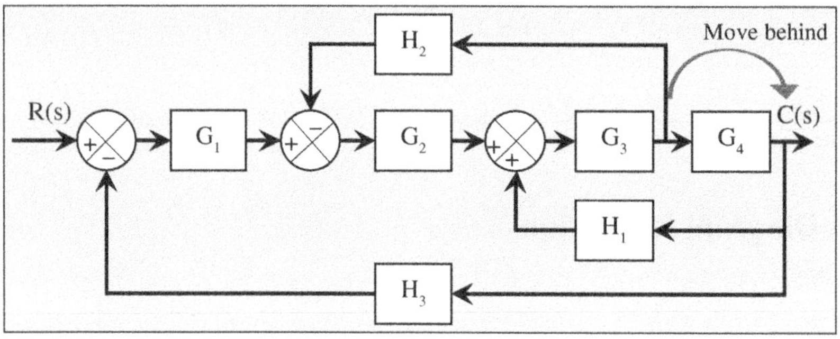

Figure 3.1 The original block diagram of Example 3.1

$$G_1 = \frac{1}{s+10},\quad G_2 = \frac{1}{s+1},\quad G_3 = \frac{s^2+1}{s^2+4s+4},\quad G_4 = \frac{s+1}{s+6},$$

$$H_1 = \frac{s+1}{s+2},\quad H_2 = 2,\quad H_3 = 1$$

Solution:

This example was a problem to solve in Ghosh [4], and it was solved in Jairth [2]. In contrast, this book shows the MATLAB steps and the transfer function for each step.

1- Start the program with "`clear`" and "`clc`" MATLAB commands.
2- Define all G's and H's block transfer functions in a polynomial form.
3- Save your file under a chosen name. Let it be EX3_1.m. The MATLAB script file will be:

```
clear
clc
% file name: EX3_1.m
nG1=[1];dG1=[1 10];
nG2=[1];dG2=[1 1];
nG3=[1 0 1];dG3=[1 4 4];
nG4=[1 1];dG4=[1 6];
nH1=[1 1];dH1=[1 2];
nH2=[2];dH2=[1];
nH3=[1];dH3=[1];
```

4- Move the takeoff point of block H_2 behind block G_4 as shown in Figs. 3.1 and 3.1-a. Use `printsys` command to show the transfer function.

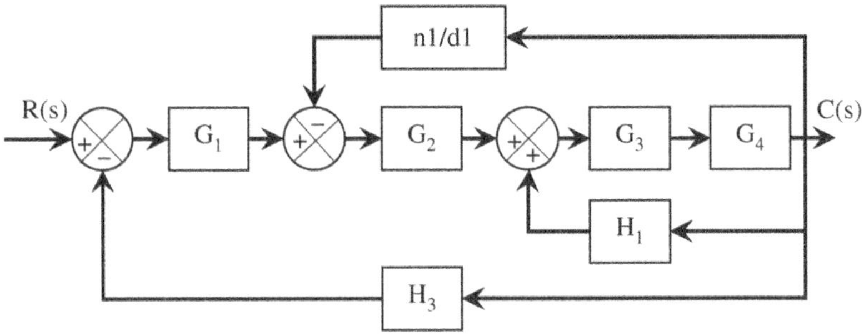

Figure 3.1-a The block diagram after completing step 4

Where $\frac{n_1}{d_1} = \frac{H_2}{G_4}$

```
n1=conv(nH2,dG4);d1=conv(dH2,nG4);
printsys(n1,d1)
```

After Execution:

```
num/den =n1/d1

   2 s + 12
   --------
     s + 1
```

5- Combining G_3 and G_4 blocks in series and eliminating the positive feedback formed by H_1 as illustrated in Fig. 3.1-a′ gives Fig. 3.1-b.

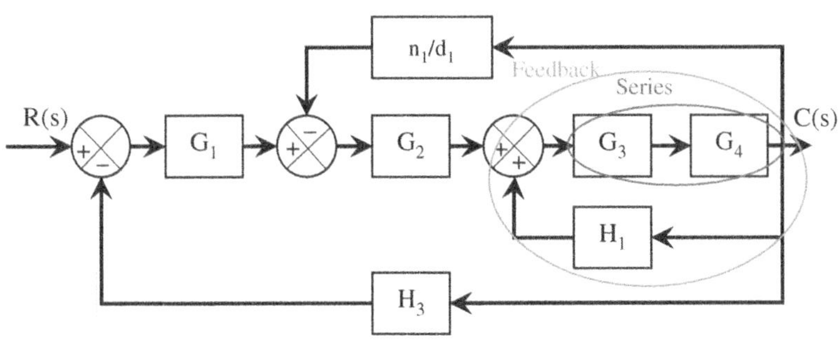

Figure 3.1-a′ Illustrations of series and feedback.

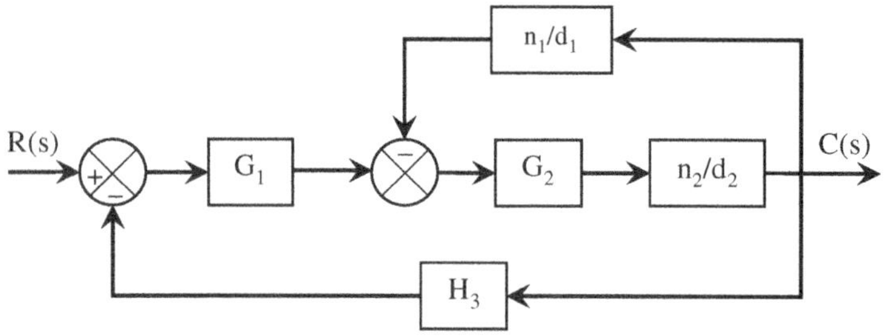

Figure 3.1-b The block diagram after completing step 5

Where $\frac{n_2}{d_2} = \frac{G_3G_4}{1 - G_3G_4H_1}$

```
[n2a,d2a]=series(nG3,dG3,nG4,dG4);
[n2,d2]=feedback(n2a,d2a,nH1,dH1,+1);
printsys(n2,d2)
```

After Execution:

```
num/den = n2/d2

   s^4 + 3 s^3 + 3 s^2 + 3 s + 2
   ------------------------------
    10 s^3 + 46 s^2 + 78 s + 47
```

6- Combining G_2 and n_2/d_2 blocks in series and eliminating the negative feedback loop formed by n_1/d_1 as demonstrated in Fig. 3.1-b′ results in Fig. 3.1-c.

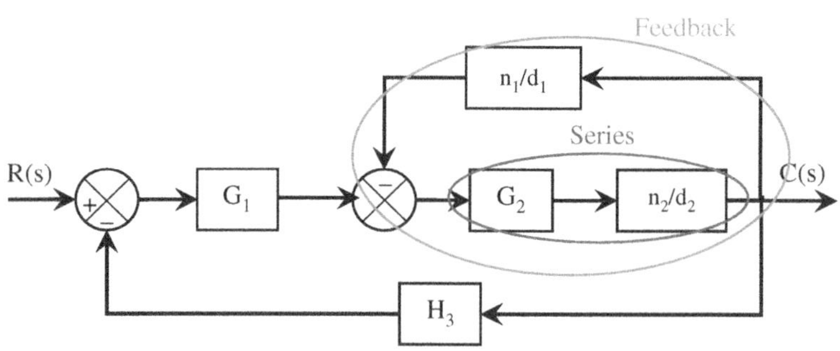

Figure 3.1-b′ Demonstrations of series and feedback.

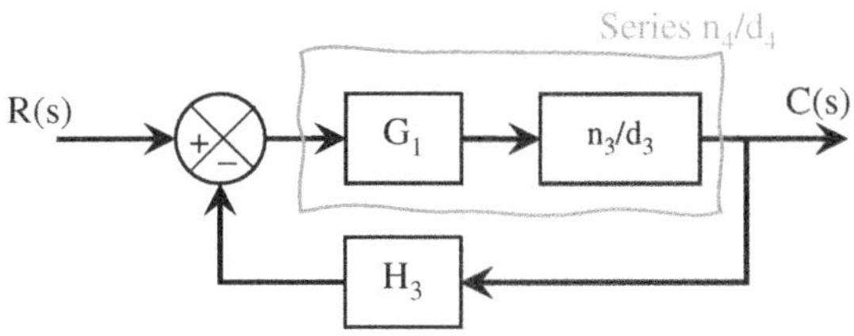

Figure 3.1-c The block diagram after completing step 6

Where $\frac{n_3}{d_3} = \frac{G_2G_3G_4}{1 - G_3G_4H_1 + G_2G_3H_2}$, $\frac{n_4}{d_4} = G_1\frac{n_3}{d_3}$

```
[n3a,d3a]=series(nG2,dG2,n2,d2);
[n3,d3]=feedback(n3a,d3a,n1,d1,-1);
 printsys(n3,d3)
```

After Execution:

```
num/den = n3/d3

          s^5 + 4 s^4 + 6 s^3 + 6 s^2 + 5 s + 2
   ---------------------------------------------------
   12 s^5 + 84 s^4 + 222 s^3 + 291 s^2 + 212 s + 71
```

7- Combining G_1 and n_3/d_3 blocks in series and eliminating the negative closed loop formed by block H_3 gives Fig. 3.1-d. The feedback formed by H_3 is called closed loop here because $H_3 = 1$.

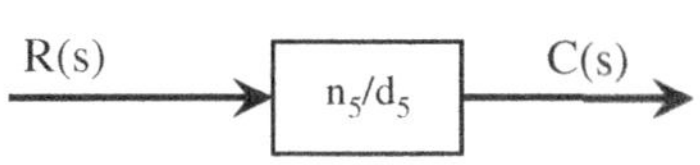

Figure 3.1-d The block diagram after completing step 7

Where $\frac{n_5}{d_5} = \frac{G_1G_2G_3G_4}{1 - G_3G_4H_1 + G_2G_3H_2 + G_1G_2G_3G_4H_3}$

```
[n4,d4]=series(nG1,dG1,n3,d3);
printsys(n4,d4)
[n5,d5]=cloop(n4,d4,-1);
printsys(n5,d5)
```

After Execution:

```
num/den = n4/d4

               s^5 + 4 s^4 + 6 s^3 + 6 s^2 + 5 s + 2
-------------------------------------------------------------
12 s^6 + 204 s^5 + 1062 s^4 + 2511 s^3 + 3122s^2 +2191s + 710
```

```
num/den = n5/d5

               s^5 + 4 s^4 + 6 s^3 + 6 s^2 + 5 s + 2
-------------------------------------------------------------
12 s^6 + 205 s^5 + 1066 s^4 + 2517 s^3 + 3128 s^2 + 2196s +712
```

Note:

Focus on the difference between `feedback` and `cloop` commands. Review Chapter 2 for more information.

For More Practice:

- See Example 13 on page 48 [1].
- See Example A-2-3 on page 47 [3].

Activity:

Write MATLAB programs for the block diagrams of those mentioned examples.

Example 3.2 Obtain C(s)/R(s) for the block diagram of Fig. 3.2 [1, p: 15, Ex. 6].

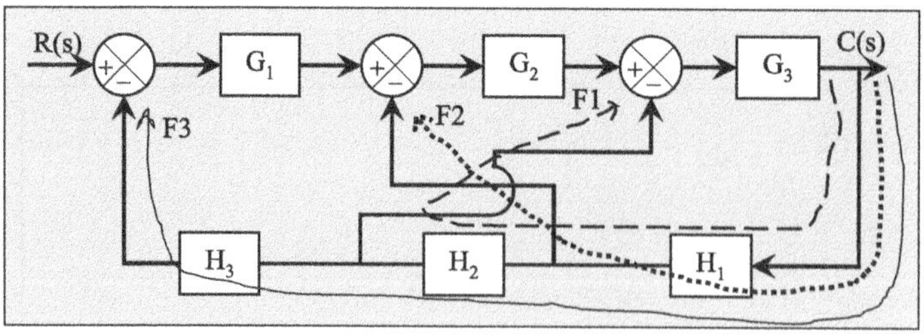

Figure 3.2 The original block diagram of Example 3.2

$$G_1 = \frac{1}{s+10},\quad G_2 = \frac{1}{s+1},\quad G_3 = \frac{s^2+1}{s^2+4s+4},$$

$$H_1 = \frac{s+1}{s+2},\quad H_2 = 2,\qquad H_3 = 1$$

Solution:

Figure 3.2 can be redrawn as shown in Fig. 3.2-a. The whole reduction procedure with the block diagrams was explained in Panda et al. [1]. This book shows the MATLAB solution with transfer functions for all G's and H's blocks. Figure 3.2 looks very confusing to students because the blocks H_1, H_2, and H_3 are drawn once time and the branches are drawn as illustrated in Fig. 3.2. The three typed lines, long dash, dot, and solid in Fig. 3.2, are illustrations for the first H_1H_2, second H_1, and third $H_1H_2H_3$ feedback loops to make Fig. 3.2-a more understandable for students.

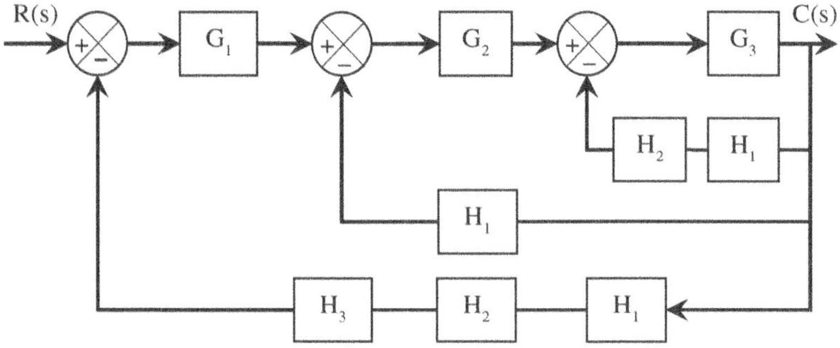

Figure 3.2-a The block diagram of Example 3.2 in more understandable form

1- Enter the numerators and denominators of all G's and H's blocks transfer functions.

```
clear
clc
% file name: EX3_2.m refer to Ex.6 page 15,Panda et. al [1]
numG1=[1]; denG1=[1  10];
numG2=[1]; denG2=[1  1];
numG3=[1  0  1]; denG3=[1  4  4];
numH1=[1  1]; denH1=[1  2];
numH2=[2]; denH2=[1];
numH3=[1]; denH3=[1];
```

2- Combine H_1 and H_2 blocks in series, then eliminate the feedback loop involving H_1H_2 as shown in Fig. 3.2-b.

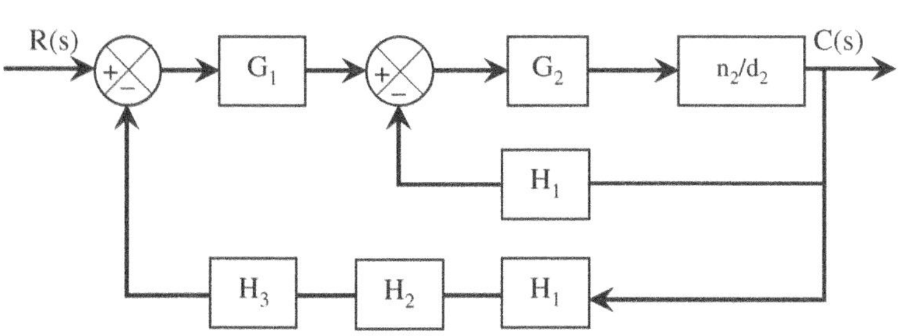

Figure 3.2-b The block diagram after completing step 2

Where $\frac{n_1}{d_1} = H_1H_2$, $\frac{n_2}{d_2} = \frac{G_3}{1 + G_3H_1H_2}$

```
[n1,d1]=series(numH1,denH1,numH2,denH2)
printsys(n1,d1)
[n2,d2]=feedback(numG3,denG3,n1,d1,-1);
printsys(n2,d2)
```

After Execution, You Will See:

```
n1 =2      2

d1 =1      2

num/den = n1/d1
   2 s + 2
   -------
    s + 2
num/den = n2/d2
     s^3 + 2 s^2 + s + 2
   ------------------------
   3s^3 + 8 s^2 + 14 s + 10
```

3- Series combination of H_1, H_2, and H_3 blocks gives Fig. 3.2-c.

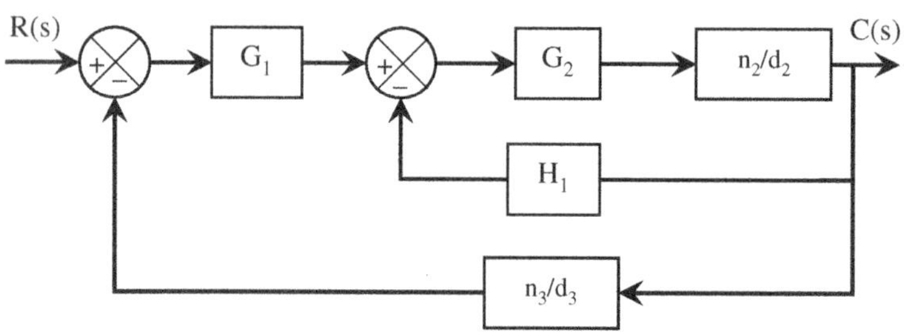

Figure 3.2-c. The block diagram after completing step 3

Where $\frac{n_3}{d_3} = H_1 H_2 H_3$

```
[n3,d3]=series(n1,d1,numH3,denH3);
printsys(n3,d3)
```

After Execution:

```
num/den = n3/d3

   2 s + 2
   -------
    s + 2
```

4- The series combination of G_2 and n_2/d_2 blocks, and the elimination of the negative feedback formed by H_1 results in Fig. 3.2-d.

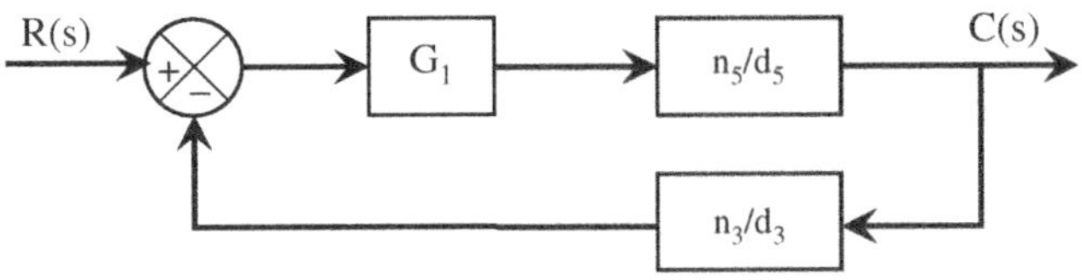

Figure 3.2-d The block diagram after completing step 4

Where $\dfrac{n_5}{d_5} = \dfrac{G_2G_3}{1 + G_3H_1H_2 + G_2G_3H_1}$

```
[n4,d4]=series(numG2,denG2,n2,d2);
printsys(n4,d4)
[n5,d5]=feedback(n4,d4,numH1,denH1,-1);
printsys(n5,d5)
```

After Execution:

```
num/den = n4/d4

                s^3 + 2 s^2 + s + 2
    -----------------------------------
    3 s^4 + 11 s^3 + 22 s^2 + 24 s + 10

num/den = n5/d5

                s^4 + 4 s^3 + 5 s^2 + 4 s + 4
    -------------------------------------------
     3s^5 + 18 s^4 + 47 s^3 + 71 s^2 + 61 s + 22
```

5- Combine G_1 and n_5/d_5 blocks in series, then eliminate the negative feedback loop involving n_3/d_3 as shown in Fig. 3.2-e.

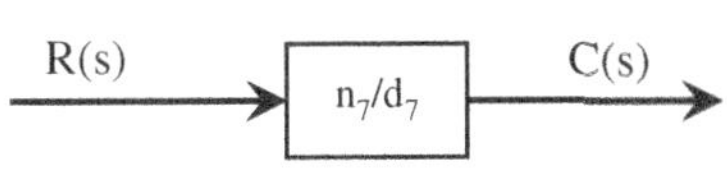

Figure 3.2-e. The block diagram after completing step 5

Where $\dfrac{n_6}{d_6} = \dfrac{G_1G_2G_3}{1 + G_3H_1H_2 + G_2G_3H_1}$,

$$\frac{n_7}{d_7} = \frac{G_1G_2G_3}{1 + G_3H_1H_2 + G_2G_3H_1 + G_1G_2G_3H_1H_2H_3}$$

```
[n6,d6]=series(numG1,denG1,n5,d5);
printsys(n6,d6)
[n7,d7]=feedback(n6,d6,n3,d3,-1);
printsys(n7,d7)
```

After Execution:

```
num/den = n6/d6

                  s^4 + 4 s^3 + 5 s^2 + 4 s + 4
   ---------------------------------------------------------
   3 s^6 + 48 s^5 + 227 s^4 + 541 s^3 + 771 s^2 + 632 s + 220
```

```
num/den = n7/d7

          s^5 + 6 s^4 + 13 s^3 + 14 s^2 + 12 s + 8
----------------------------------------------------------------
3 s^7 + 54 s^6 + 325 s^5 + 1005 s^4 + 1871 s^3 + 2192 s^2 +
1500 s + 448
```

`printsys` command was used after each reduction step to show the transfer function of that step.

Example 3.3 Simplify the block diagram of Fig. 3.3 [1, p: 14, Ex. 5].

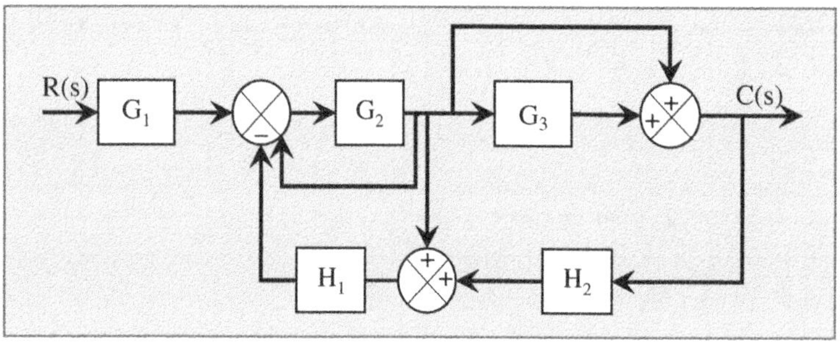

Figure 3.3 The original block diagram of Example 3.3

$$G_1 = \frac{10}{s^2+2s},\quad G_2 = \frac{s+1}{s+2},\quad G_3 = \frac{2}{s+1},\quad H_1 = \frac{6s}{s+1},\quad H_2 = \frac{s}{s+1}$$

Solution:

The reduction process and the overall transfer function C(s)/R(s) in terms of G's and H's were presented in Panda et al. [1]. This book presents two step-by-step programming strategies to get the overall transfer functions using MATLAB commands.

In the first code, the numerators and denominators for each block transfer function are defined separately. `Printsys` command was used in the first code to present the transfer function after each reduction process in fraction form. On the other hand, the `s=tf('s')` command was used in the second code to deal with the transfer function in fraction form. So, for this reason, there was no need to use the `printsys` command here.

The First Code:

1- Write "`clear`" and "`clc`" to delete any previous execution.
2- Define all G's and H's numerators and denominators.

```
clear
clc
% file name: Ex3_3_1.m refer to ex.5 p:14, [1]
nG1=[10];    dG1=[1  2  0];
nG2=[1  1]; dG2=[1  2];
nG3=[2];     dG3=[1  1];
nH1=[6  0]; dH1=[1  1];
nH2=[1  0]; dH2=[1  1];
```

3- The eliminations of the miner feedforward path involving +1 and the miner closed loop of -1 of Fig. 3.3 result in Fig. 3.3-a.

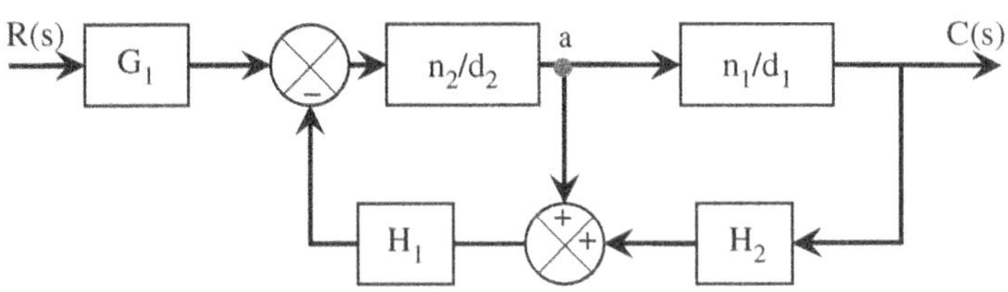

Figure 3.3-a The block diagram after completing step 3

Where $\frac{n_1}{d_1} = G_3 + 1, \ \frac{n_2}{d_2} = \frac{G_2}{1 + G_2}$

```
[n1,d1]=parallel(nG3,dG3,1,1)
printsys(n1,d1)
[n2,d2]=cloop(nG2,dG2,-1);
printsys(n2,d2)
```

After Execution, The Result Will Be:

```
n1 =
       1     3
d1 =
       1     1

num/den = n1/d1

       s + 3
       -----
       s + 1

num/den = n2/d2
       s + 1
      -------
      2 s + 3
```

4- Moving the takeoff point "a" that is placed between n_2/d_2 and n_1/d_1 behind the block n_1/d_1 as shown in Figs. 3.3-a and 3.3-b.

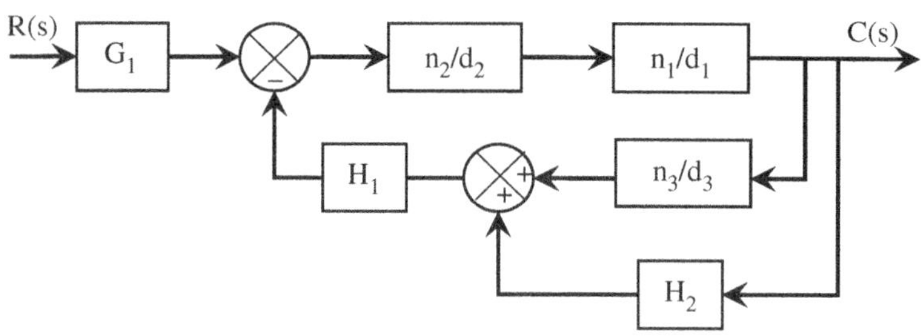

Figure 3.3-b The block diagram after completing step 4

Where $\frac{n_3}{d_3} = \frac{1}{\frac{n_1}{d_1}}$

```
% moving takeoff point after a block, 1/(tf(n1,d1))
n3=d1;
d3=n1;
printsys(n3,d3)
```

After Execution, The Result Will Be:

```
num/den = n3/d3

    s + 1
    -----
    s + 3
```

5- Combining n_3/d_3 and H_2 blocks in parallel and combining n_2/d_2 and n_1/d_1 in series give Fig. 3.3-c

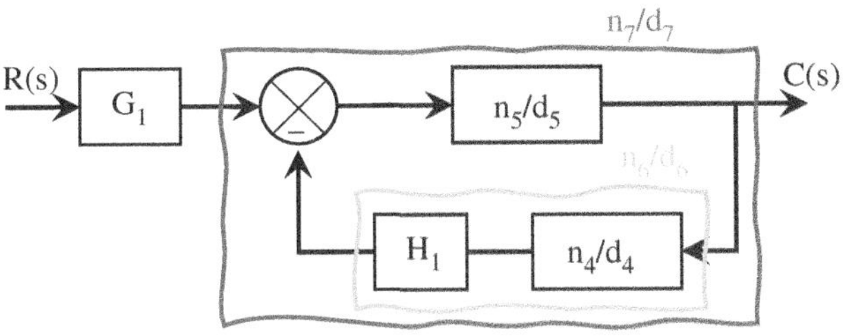

Figure 3.3-c The block diagram after completing step 5

Where $\frac{n_5}{d_5} = \frac{n_2}{d_2} * \frac{n_1}{d_1}$, $\frac{n_4}{d_4} = \frac{n_3}{d_3} + H_2$, $\frac{n_6}{d_6} = H_1 * \frac{n_4}{d_4}$

$$\frac{n_7}{d_7} = \frac{G_2 + G_2G_3}{1 + G_2 + G_2H_1 + G_2H_1H_2 + G_2 + G_2G_3H_1H_2}$$

```
[n4,d4]=parallel(n3,d3,nH2,dH2);
printsys(n4,d4)
[n5,d5]=series(n2,d2,n1,d1);
printsys(n5,d5)
```

After Execution, The Result Will Be:

```
num/den = n4/d4

   2 s^2 + 5 s + 1
   ---------------
    s^2 + 4 s + 3

num/den = n5/d5
    s^2 + 4 s + 3
   ---------------
   2 s^2 + 5 s + 3
```

6- Combining H_1 and n_4/d_4 blocks in series gives n_6/d_6. Eliminating the negative feedback including n_6/d_6 gives n_7/d_7. Combining G_1 with n_7/d_7 gives n_8/d_8 as shown in Fig. 3.3-d

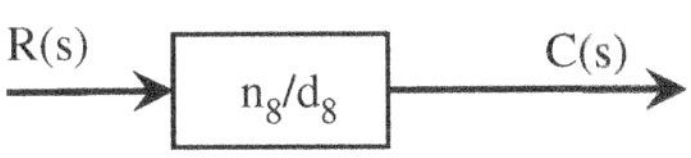

Figure 3.3-d The block diagram after completing step 6

Where $\frac{n_8}{d_8} = \frac{G_1(G_2 + G_2G_3)}{1 + G_2 + G_2H_1 + G_2H_1H_2 + G_2 + G_2G_3H_1H_2}$

```
[n6,d6]=series(n4,d4,nH1,dH1);
printsys(n6,d6)
[n7,d7]=feedback(n5,d5,n6,d6,-1);
printsys(n7,d7)
[n8,d8]=series(nG1,dG1,n7,d7);
printsys(n8,d8)
```

After Execution, The Result Will Be:

```
num/den = n6/d6

    12 s^3 + 30 s^2 + 6 s
   ----------------------
    s^3 + 5 s^2 + 7 s + 3

num/den = n7/d7

      s^5 + 9 s^4 + 30 s^3 + 46 s^2 + 33 s + 9
   ---------------------------------------------------
   14 s^5 + 93 s^4 + 204 s^3 + 170 s^2 + 54 s + 9
```

```
num/den = n8/d8

         10 s^5 + 90 s^4 + 300 s^3 + 460 s^2 + 330 s + 90
----------------------------------------------------------------------
14 s^7 + 121 s^6 + 390 s^5 + 578 s^4 + 394 s^3 + 117 s^2 + 18 s
```

The Second Code:

```
clear
clc
% % file name: Ex3_3_2.m refer to ex.5 p:14, [1]
% Another programming procedure
s=tf('s');
G1=10/(s^2+2*s);
G2=(s+1)/(s+2);
G3=2/(s+1);
H1=6*s/(s+1);
```

```
  H2=s/(s+1);
  sys1=parallel(G3,1)
% Use feedback instead of cloop
sys2=feedback(G2,1,-1)
% Moving takeoff point after a block 1/sys1
sys3=1/sys1
sys4=parallel(sys3,H2)
sys5=series(sys2,sys1)
sys6=series(sys4,H1)
sys7=feedback(sys5,sys6,-1)
sys8=series(G1,sys7)
```

After Execution:

```
sys1 =
  s + 3
  -----
  s + 1

sys2 =
   s + 1
  --------
  2 s + 3

sys3 =
  s + 1
  -----
  s + 3

sys4 =
  2 s^2 + 5 s + 1
  ----------------
   s^2 + 4 s + 3

sys5 =
   s^2 + 4 s + 3
  ----------------
  2 s^2 + 5 s + 3

sys6 =
  12 s^3 + 30 s^2 + 6 s
  -----------------------
  s^3 + 5 s^2 + 7 s + 3
```

```
sys7 =
     s^5 + 9 s^4 + 30 s^3 + 46 s^2 + 33 s + 9
  ------------------------------------------------
  14 s^5 + 93 s^4 + 204 s^3 + 170 s^2 + 54 s + 9
```

```
sys8 =
          10 s^5 + 90 s^4 + 300 s^3 + 460 s^2 + 330 s + 90
  ---------------------------------------------------------------------
  14 s^7 + 121 s^6 + 390 s^5 + 578 s^4 + 394 s^3 + 117 s^2 + 18 s
```

Note that the same overall transfer function is acquired from the two programs.

Example 3.4 Obtain the overall Transfer function for the control system of Fig. 3.4.

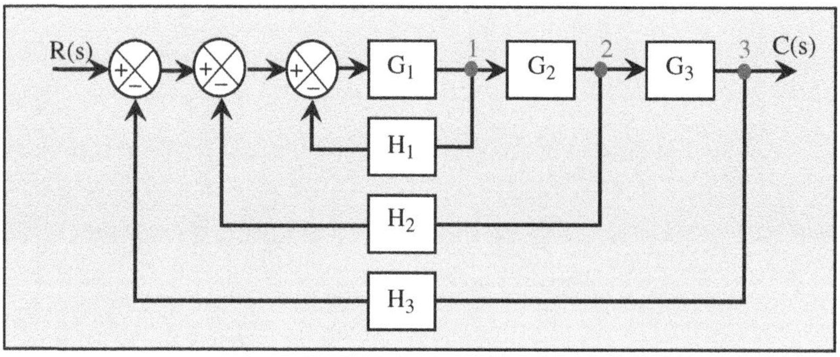

Figure 3.4 The original block diagram of Example 3.4.

Solution:

This example can be solved in more than three different ways. Here we will discuss two and leave the last one to students for more practice.

The First Method:

We will call this method the blind method because most students will follow it, and it doesn't require a good knowledge of block diagram reduction rules. You have three feedback loops as illustrated in Fig. 3.4-a. Eliminate each feedback loop, and then the result will be in series with the following block. Repeat the process for the other two feedback loops.

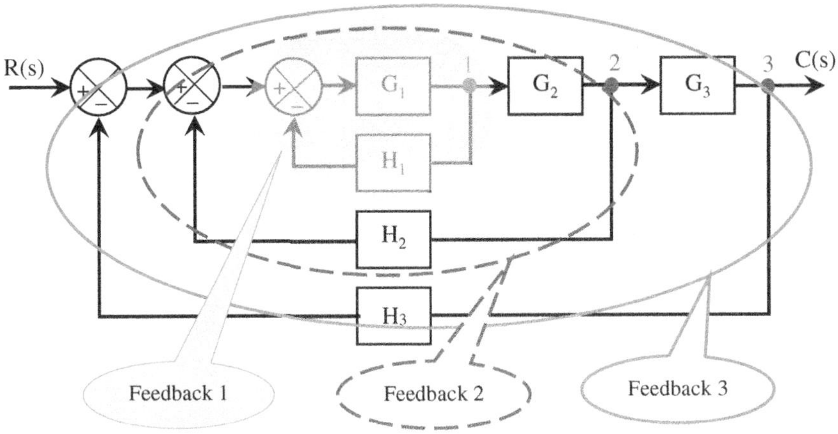

Figure 3.4-a Feedback loops illustration.

The overall transfer function will be:

$$\frac{C(s)}{R(s)} = \frac{G_1G_2G_3}{1 + G_1H_1 + G_1G_2H_2 + G_1G_2G_3H_3}$$

The pseudocode will be:

1- `Clear.`
2- `clc.`
3- Define numerators and denominators for all G's and H's blocks.
4- Eliminate the negative feedback (G_1, H_1).
5- Series (result of step 4, G_2).
6- Eliminate the negative feedback (result of step 5, H_2).
7- Series (result of step 6, G_3).
8- Eliminate the negative feedback (results of step 7, H_3).

The Second Method:

1- Moving the takeoff point "1" of block H_1 after the block G_2 as demonstrated as the first move in Fig. 3.4-b.
2- Moving the takeoff point "2" of block H_2 after the block G_3 as demonstrated as the second move in Fig. 3.4-b.

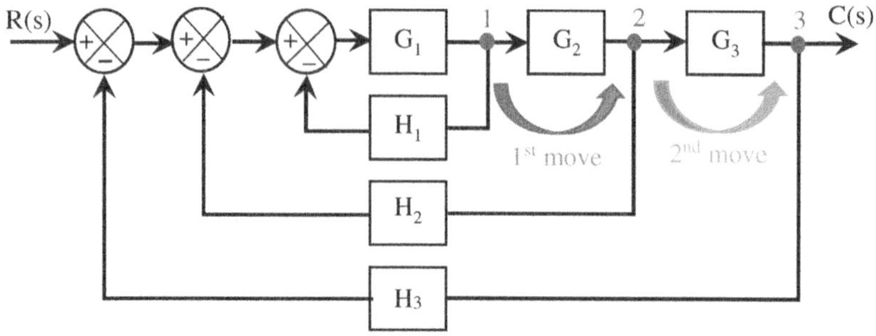

Figure 3.4-b The block diagram demonstration of steps 1 and 2.

Doing steps 1 and 2 results in Fig. 3.4-c

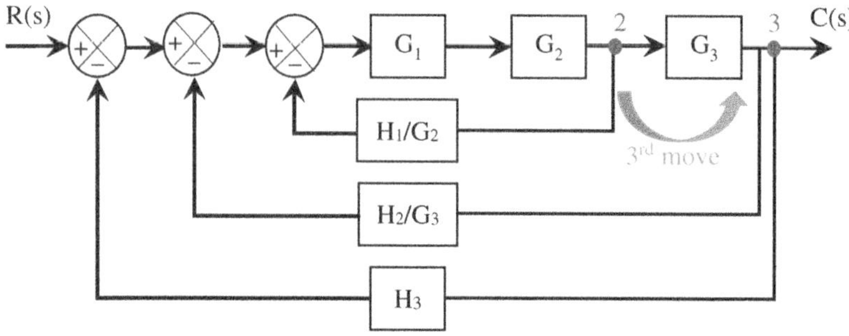

Figure 3.4-c The block diagram after completing steps 1 and 2 and demonstration of step 3.

3- Moving the takeoff point "2" of block H_1/G_2 after block G_3 as demonstrated as the third move in Fig. 3.4-c gives in Fig. 3.4-d.

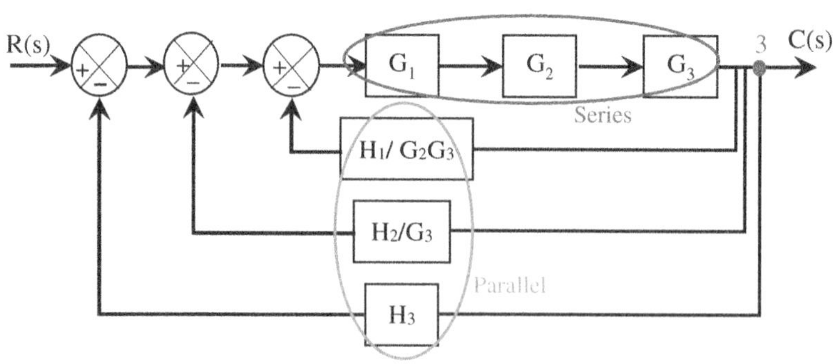

Figure 3.4-d The block diagram after completing step 3.

4- Combining the series group of G_1, G_2, and G_3 blocks and the parallel group of H_1/G_2G_3, H_2/G_3, and H_3 blocks results in Fig. 3.4-e.

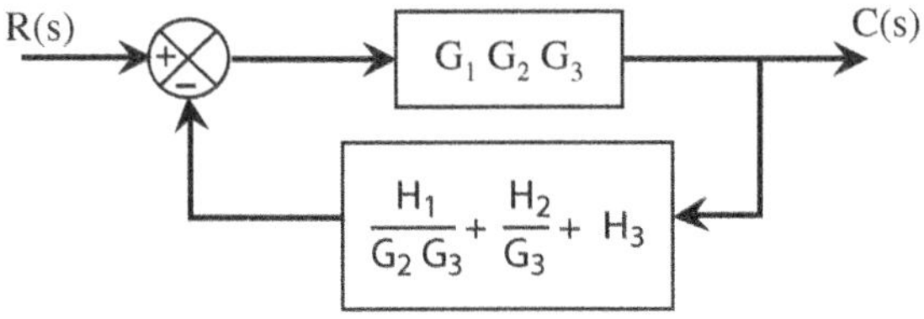

Figure 3.4-e The block diagram after completing step 4.

5- Eliminating the feedback loop formed by $\frac{H_1}{G_2G_3} + \frac{H_2}{G_3} + H_3$ gives Fig. 3.4-f.

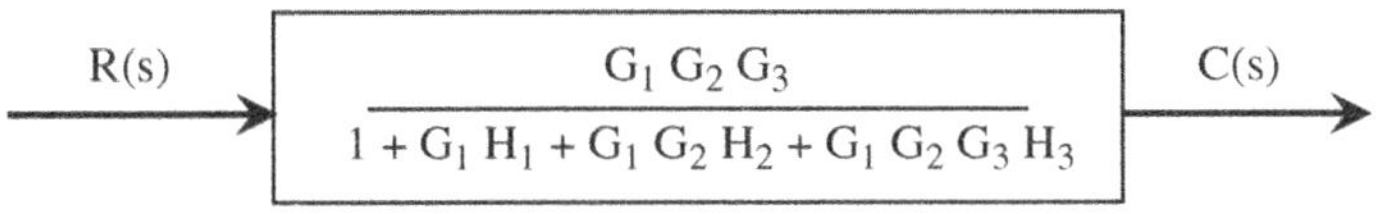

Figure 3.4-f The block diagram after completing step 5.

See how the overall transfer function is the same in the two procedures.

$$\frac{C(s)}{R(s)} = \frac{G_1G_2G_3}{1+G_1H_1+G_1G_2H_2+G_1G_2G_3H_3}$$

Activity:

- Reduce the block diagram differently.
- Try to make a MATLAB code, but first, you have to assume transfer functions for all the G's and H's blocks.

For More Practice:

See Chapter 6, block diagram on pages (141–176) [4].

Example 3.5 Obtain C(s)/R(s) for the block diagram of Fig. 3.5 [3, p: 27, Ex. 2-1].

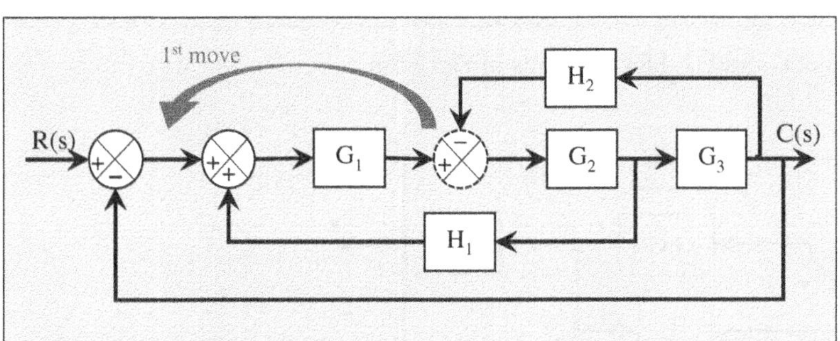

Figure 3.5 The original block diagram of Example 3.5.

Solution:

This example had been solved in the mentioned reference, but here we will follow a different method, in addition to a pseudocode.

The block diagram shown in Fig. 3.5 contains three summing points:

- Summing point of the negative closed loop.
- Summing point of the positive feedback loop formed by H_1.
- Summing point of the negative feedback loop formed by H_2.

1- Start with "`clear`" and "`clc`", then enter numerators and denominators for all G's and H's blocks. The pseudocode will be:

> 1- Use `clear` to remove and clean any previous program execution.
> 2- Use `clc` to clean your screen.
> 3- Add a note like, % file name: Ex 3.5.m refer to example 2-1, page 27 in Ogata [3].
> 4- Define numerators and denominators for all G's and H's blocks. For example, if $G_1 = 1/(s+10)$, write `ng1=[1];dg1=[1 10];`
> 5- Continue following the same process for the rest numerators and denominators nG_2, dG_2, nG_3, dG_3, nH_1, dH_1, nH_2, and dH_2.

2- Swapping the third summing point before the second one as demonstrated by the first move in Fig. 3.5 gives Fig. 3.5-a.

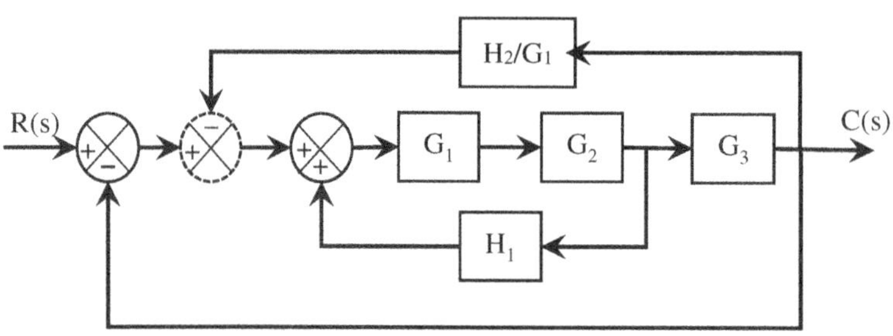

Figure 3.5-a The block diagram after completing step 2.

> 6- Use `conv` command to get H_2/G_1
> `n1=conv(nH2,dG1), d1=conv(dH2,nG1)`

3- The series combination of G_1 and G_2 blocks results in Fig. 3.5-b.

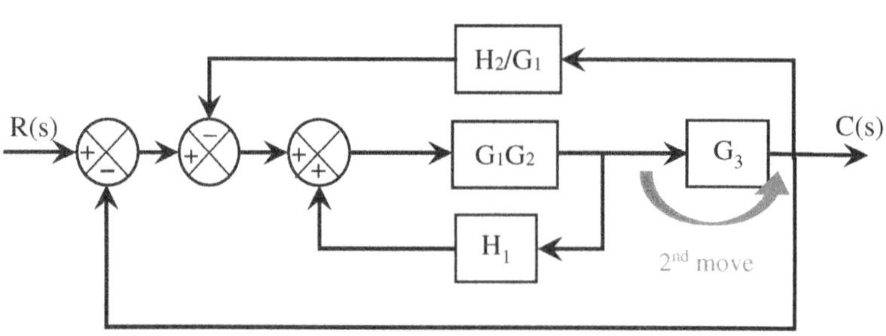

Figure 3.5-b The block diagram after completing step 3.

```
7- Series (G1 and G2);
```

4- Moving the takeoff point of block H_1 after block G_3 as demonstrated as the second move in Fig. 3.5-b gives Fig. 3.5-c.

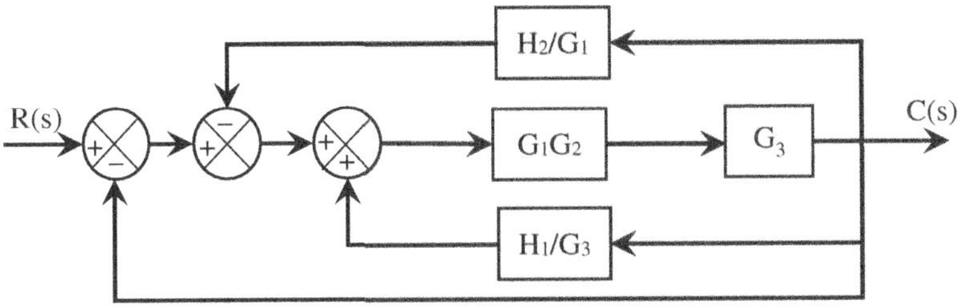

Figure 3.5-c The block diagram after completing step 4.

```
8- Use conv command to get H1/G3
  n3=conv(nH1,dG3); d3=conv(dH1,nG3);
```

5- Combine G_1G_2 and G_3 blocks in series, then combine the blocks H_2/G_1 and H_1/G_3 in parallel to get Fig. 3.5-d.

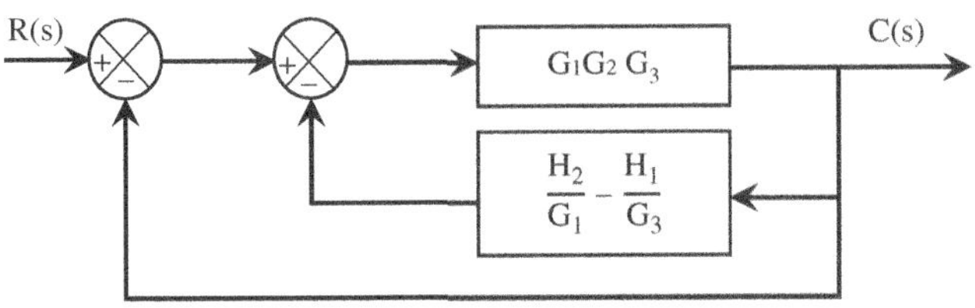

Figure 3.5-d The block diagram after completing step 5.

```
9- Series (G1G2 and G3);
10- Parallel (H2/G1 and H1/G3);
```

6- Eliminate the negative feedback formed by $\frac{H_2}{G_1} - \frac{H_1}{G_3}$ to get Fig. 3.5-e.

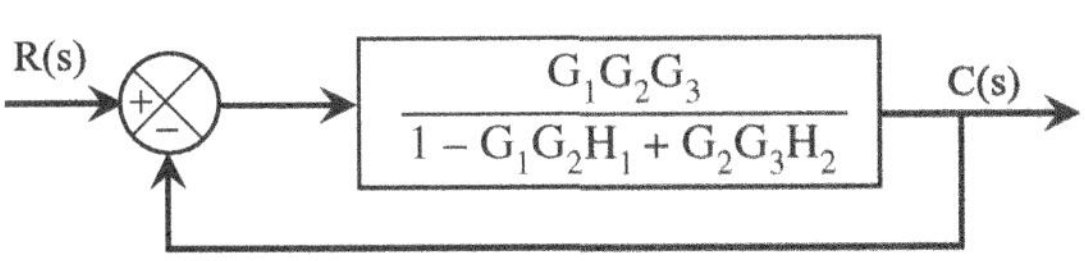

Figure 3.5-e The block diagram after completing step 6.

11- Feedback $\left(G_1G_2G_3 \text{ and } \frac{H_2}{G_1} - \frac{H_1}{G_3}\right)$

7- Finally, eliminate the closed loop as shown in Fig. 3.5-f.

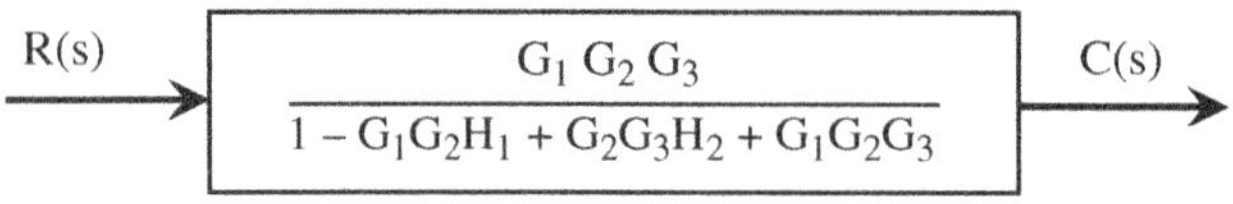

Figure 3.5-f The block diagram after completing step 7.

> 12- Cloop $\left(\frac{G_1G_2G_3}{1 - G_1G_2H_1 + G_2G_3H_2} \text{ and } -1\right)$;
>
> 13- Use printsys to get the overall transfer function.

See, we used the "`printsys`" command at the end to get the final transfer function.

Activity:

Write a MATLAB script file for Example 3.5 pseudocode.

Example 3.6 Simplify the block diagram of Fig. 3.6.

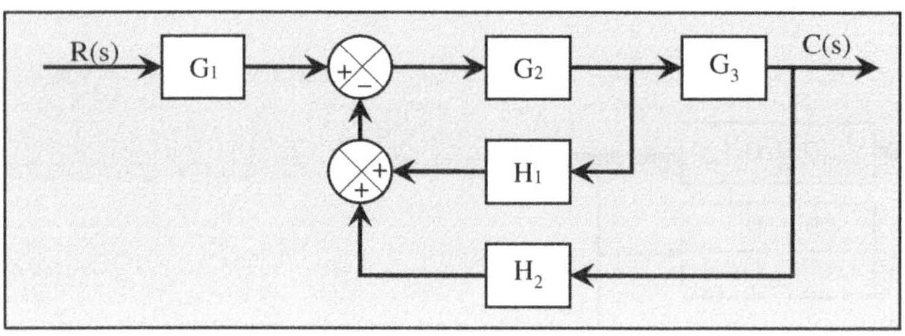

Figure 3.6 The original block diagram of Example 3.6

$$G_1 = \frac{10}{s^2 + 2s},\; G_2 = \frac{s+1}{s+2},\; G_3 = \frac{2}{s+1},\; H_1 = \frac{6s}{s+1},\; H_2 = \frac{s}{s+1}$$

Solution:

1- Figure 3.6 can be redrawn as in Fig. 3.6-a.

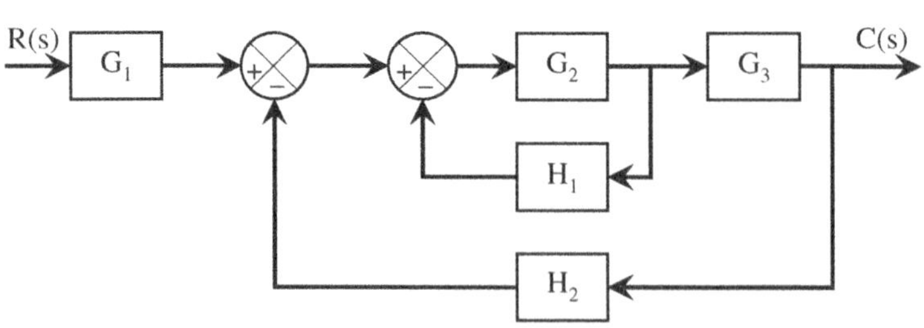

Figure 3.6-a Another way to draw the block diagram of Fig. 3.6.

2- Start the program with "`clear`" and "`clc`", then enter all G's and H's transfer functions.

```
clear
clc
% file name: Ex 3_6.m
s=tf('s');
G1=10/(s^2+2*s);
G2=(s+1)/(s+2);
G3=2/(s+1);
H1=6*s/(s+1);
H2=s/(s+1);
```

3- Eliminating the minor negative feedback formed by H_1 gives Fig. 3.6-b.

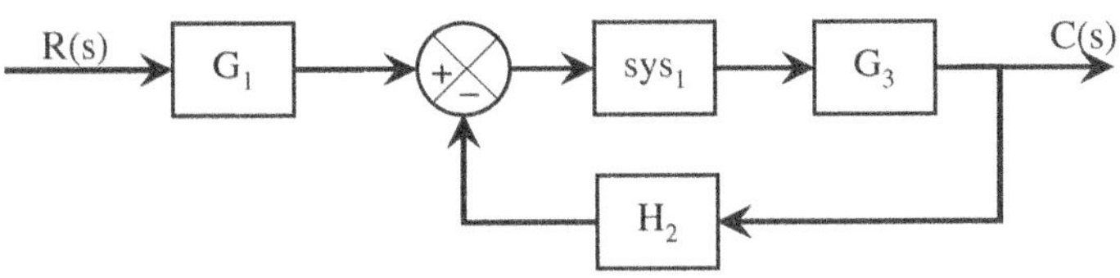

Figure 3.6-b The block diagram after completing step 3

Where $sys_1 = \frac{G_2}{1 + G_2H_1}$

```
sys1=feedback(G2,H1,-1)
```

After Execution:

```
sys1 =

   s^2 + 2 s + 1
  -----------------
  7 s^2 + 9 s + 2
```

4- Combining sys_1 and G_3 blocks in series gives Fig. 3.6-c.

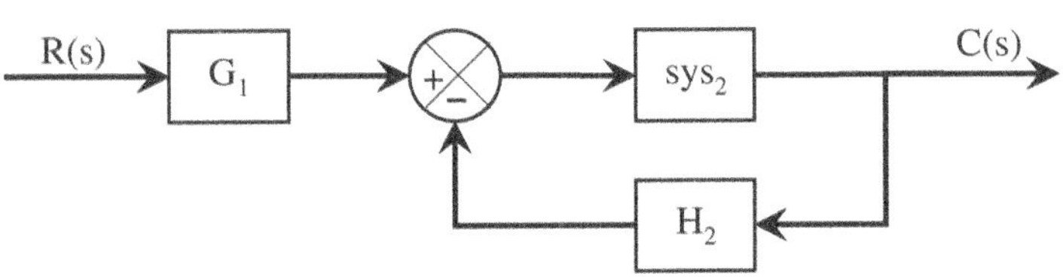

Figure 3.6-c The block diagram after completing step 4

Where $sys_2 = sys_1G_3$

```
11- sys2=series(sys1,G3)
```

After Execution:

```
sys2 =

         2 s^2 + 4 s + 2
  ---------------------------
  7 s^3 + 16 s^2 + 11 s + 2
```

5- Eliminating the negative feedback involving H_2 gives Fig. 3.6-d.

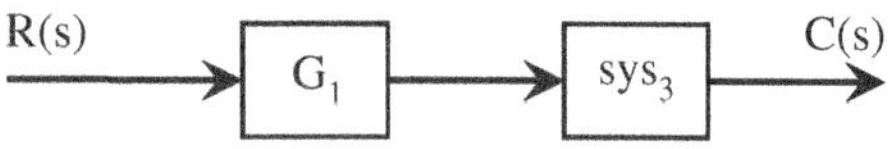

Figure 3.6-d The block diagram after completing step 5

Where $sys_3 = \dfrac{sys_2}{1 + sys_2 H_2}$

```
12- sys3=feedback(sys2,H2,-1)
```

After Execution:

```
sys3 =

         2 s^3 + 6 s^2 + 6 s + 2
  -----------------------------------
  7 s^4 + 25 s^3 + 31 s^2 + 15 s + 2
```

6- Combining G_1 and sys_3 blocks in series results in Fig. 3.6-e.

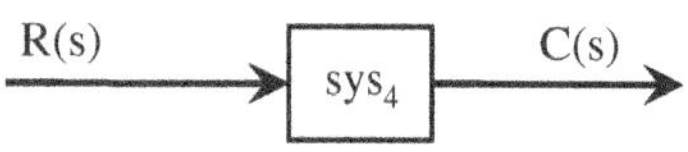

Figure 3.6-e The block diagram after completing step 6

Where $sys_4 = G_1 sys_3$

```
13- sys4=series(G1,sys3)
```

After Execution:

```
sys4 =
             20 s^3 + 60 s^2 + 60 s + 20
  ---------------------------------------------------
  7 s^6 + 39 s^5 + 81 s^4 + 77 s^3 + 32 s^2 + 4 s
```

Note there is no need to use the "`printsys`" command here because we are dealing with the whole transfer function like "`sys1, sys2, sys3` and `sys4`".

For More Practice:

See Ex. 14 on page 50 [1].

Example 3.7 Simplify the block diagram of Fig. 3.7. This figure is closer to Ogata [3, p: 46, Ex. A-2-2].

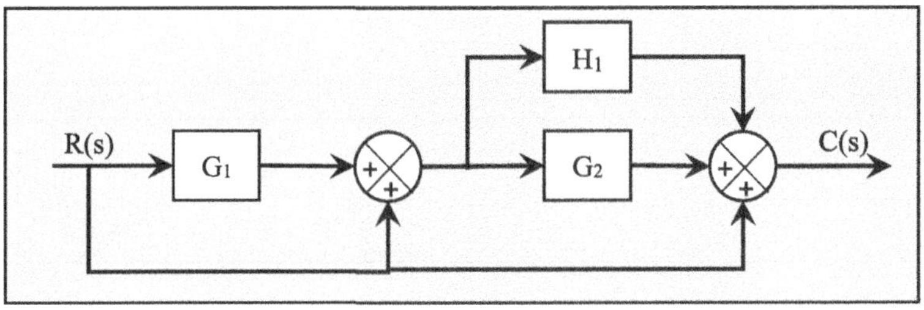

Figure 3.7 The original block diagram of Example 3.7

$$G_1 = \frac{10}{s^2 + 2s},\ G_2 = \frac{s+1}{s+2},\ H_1 = \frac{6s}{s+1}$$

Solution:

1- Replot Fig. 3.7 as shown in Fig. 3.7-a.

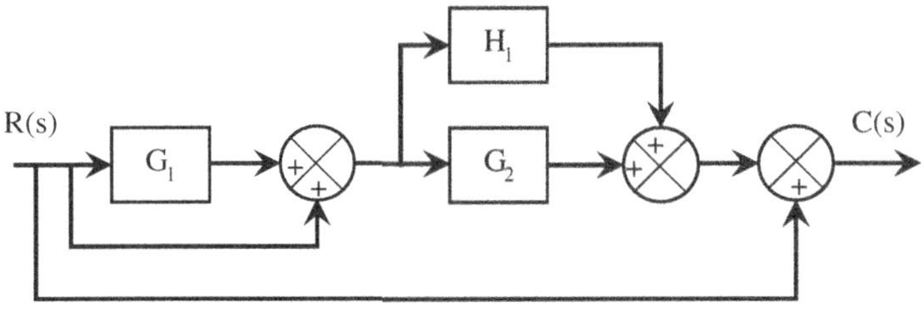

Figure 3.7-a Another way to draw the block diagram of Fig. 3.7.

2- Start the script file by writing "`clear`" and "`clc`", and enter all G's and H as transfer functions.

```
clear
clc
% file name: Ex 3_7.m
s=tf('s');
G1=10/(s^2+2*s);
G2=(s+1)/(s+2);
H1=6*s/(s+1);
```

3- Eliminating the two feedforward paths involving H_1 and 1 gives Fig. 3.7-b.

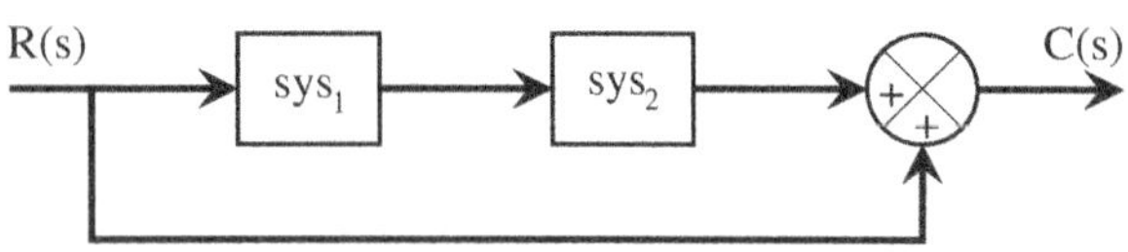

Figure 3.7-b The block diagram after completing step 3

Where $sys_1 = G_1 + 1,\qquad sys_2 = H_1 + G_2$

```
sys1=parallel(G1,1)
sys2=parallel(G2,H1)
```

After Execution:

```
sys1 =

  s^2 + 2 s + 10
  --------------
    s^2 + 2 s

sys2 =

  7 s^2 + 14 s + 1
  ----------------
   s^2 + 3 s + 2
```

4- Combine sys_1 and sys_2 blocks in series, then eliminate the feedforward path of 1 to get Fig. 3.7-c.

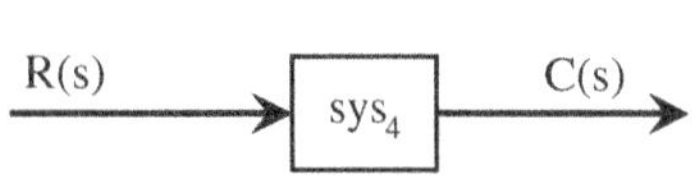

Figure 3.7-c The block diagram after completing step 4

Where $sys_3 = (G_1 + 1)(H_1 + G_2),\qquad sys_4 = (G_1 + 1)(H_1 + G_2) + 1$

```
10- sys3=series(sys1,sys2)
11- sys4=parallel(sys3,1)
```

After Execution:

```
sys3 =
   7 s^4 + 28 s^3 + 99 s^2 + 142 s + 10
  -------------------------------------
        s^4 + 5 s^3 + 8 s^2 + 4 s
```

```
sys4 =

  8 s^4 + 33 s^3 + 107 s^2 + 146 s + 10
  --------------------------------------
       s^4 + 5 s^3 + 8 s^2 + 4 s
```

For More Practice:

See Examples A2-1 and A2-2 on page 46 [3]. They are approximately similar to this example.

Example 3.8 Simplify the block diagram of Fig. 3.8 [3, p: 60, Problem B 2-3].

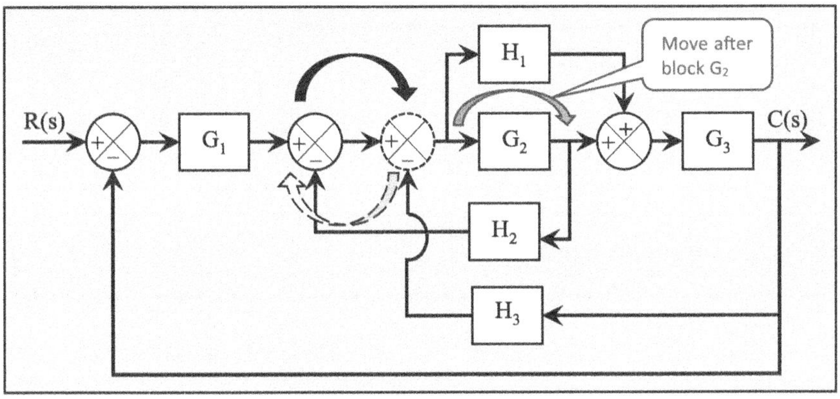

Figure 3.8 The original block diagram of Example 3.8

$$G_1 = \frac{10}{s+2},\ G_2 = \frac{s+1}{s+2},\ G_3 = \frac{2}{s+1},\ H_1 = \frac{6s}{s+1},\ H_2 = \frac{1}{s+1},\ H_3 = \frac{2}{s+2}$$

Solution:

1- Start the script file with "`clear`" and "`clc`", and enter all G's and H's numerators and denominators.

```
clear
clc
% file name: Ex 3_8.m
nG1=[10]; dG1=[1  2];
nG2=[1  1]; dG2=[1  2];
nG3=[2]; dG3=[1  1];
nH1=[6  0]; dH1=[1  1];
nH2=[1]; dH2=[1  1];
nH3=[2]; dH3=[1  2];
```

2- By rearranging the second and the third summing points, and moving the takeoff point of block H_1 after G_2. The block diagram of Fig. 3.8 will be shown in Fig. 3.8-a.

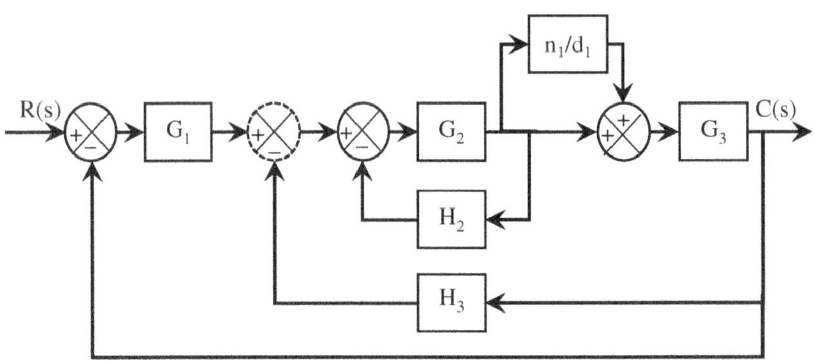

Figure 3.8-a The block diagram after completing step 2

Where $\frac{n_1}{d_1} = \frac{H_1}{G_2}$

```
n1=conv(nH1,dG2), d1=conv(dH1,nG2)
```

After Execution:

```
n1 =
      6    12     0

d1 =
      1     2     1
```

3- Eliminating the two loops involving n_1/d_1 and H_2 results in Fig. 3.8-b.

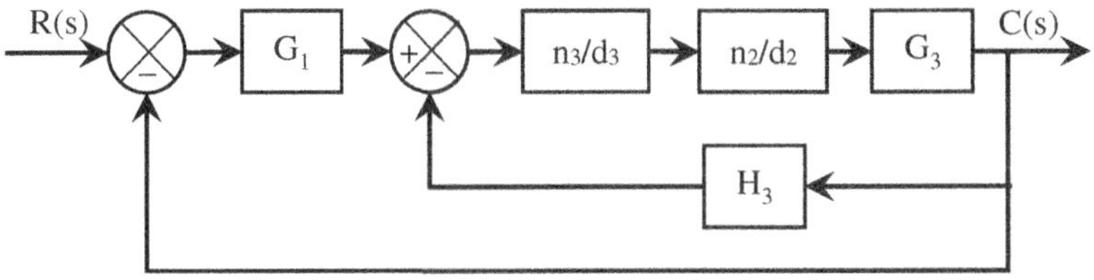

Figure 3.8-b The block diagram after completing step 3

Where $\frac{n_2}{d_2} = \frac{H_1}{G_2} + 1, \quad \frac{n_3}{d_3} = \frac{G_2}{1 + G_2H_2}$

```
11- [n2,d2]=parallel(n1,d1,1,1);
12- printsys(n2,d2)
13-  [n3,d3]=feedback(nG2,dG2,nH2,dH2,-1);
14- printsys(n3,d3)
```

After Execution:

```
num/den = n2/d2

   7 s^2 + 14 s + 1
   ----------------
     s^2 + 2 s + 1

num/den = n3/d3

   s^2 + 2 s + 1
   --------------
   s^2 + 4 s + 3
```

4- Combining n_2/d_2, n_3/d_3, and G_3 blocks in series gives Fig. 3.8-c.

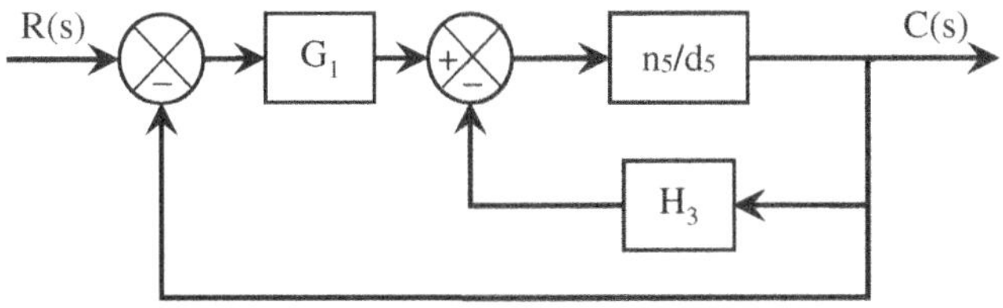

Figure 3.8-c The block diagram after completing step 4

Where $\frac{n_4}{d_4} = \frac{H_1 + G_2}{1 + G_2H_2}, \quad \frac{n_5}{d_5} = \frac{(H_1 + G_2)G_3}{1 + G_2H_2}$

```
[n4,d4]=series(n3,d3,n2,d2);
[n5,d5]=series(n4,d4,nG3,dG3);
printsys(n5,d5)
```

After Execution:

```
num/den = n5/d5

     14 s^4 + 56 s^3 + 72 s^2 + 32 s + 2
   ----------------------------------------------
   s^5 + 7 s^4 + 18 s^3 + 22 s^2 + 13 s + 3
```

5- Eliminating the feedback loop formed by H_3 results in Fig. 3.8-d.

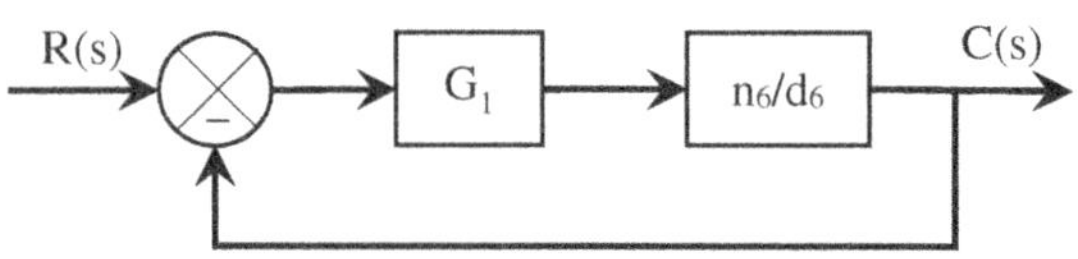

Figure 3.8-d The block diagram after completing step 5

Where $\frac{n_6}{d_6} = \frac{(H_1 + G_2)G_3}{1 + G_2H_2 + (H_1 + G_2)G_3H_3}$

```
[n6,d6]=feedback(n5,d5,nH3,dH3,-1);
printsys(n6,d6)
```

After Execution:

```
num/den = n6/d6

     14 s^5 + 84 s^4 + 184 s^3 + 176 s^2 + 66 s + 4
   ------------------------------------------------------------
   s^6 + 9 s^5 + 60 s^4 + 170 s^3 + 201 s^2 + 93 s + 10
```

6- The series combination of G_1 and n_6/d_6 blocks gives Fig. 3.8-e

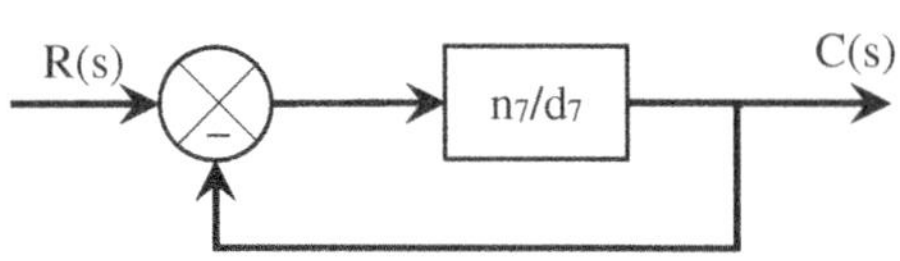

Figure 3.8-e The block diagram after completing step 6

Where $\frac{n_7}{d_7} = \frac{(H_1 + G_2)G_3G_1}{1 + G_2H_2 + (H_1 + G_2)G_3H_3}$

```
20- [n7,d7]=series(n6,d6,nG1,dG1);
21- printsys(n7,d7)
```

After Execution:

```
num/den = n7/d7

     140 s^5 + 840 s^4 + 1840 s^3 + 1760 s^2 + 660 s + 40
 -------------------------------------------------------------
 s^7 + 11 s^6 + 78 s^5 + 290 s^4 +541 s^3 +495 s^2 +196 s+ 20
```

7- Finally, eliminating the closed loop results in Fig. 3.8-f

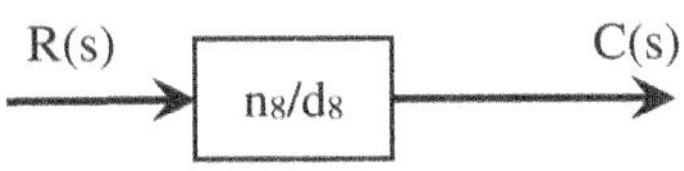

Figure 3.8-f The block diagram after completing step 7

Where $\dfrac{n_8}{d_8} = \dfrac{(H_1 + G_2)G_3G_1}{1 + G_2H_2 + (H_1 + G_2)G_3H_3 + (H_1 + G_2)G_3G_1}$

```
22- [n8,d8]=cloop(n7,d7,-1);
23- printsys(n8,d8)
```

After Execution:

```
num/den = n8/d8

     140 s^5 + 840 s^4 + 1840 s^3 + 1760 s^2 + 660 s + 40
 -------------------------------------------------------------
s^7 + 11 s^6 + 218 s^5 + 1130 s^4 + 2381 s^3 + 2255 s^2 + 856
s+ 60
```

For More Practice:

See also Example 13 on page 48 [1].

Example 3.9 Obtain C(s)/R(s) and C(s)/D(s) transfer functions for the two-input, one-output system of Fig. 3.9. This figure is closer to Ogata [3, p: 48, ex. A-2-4].

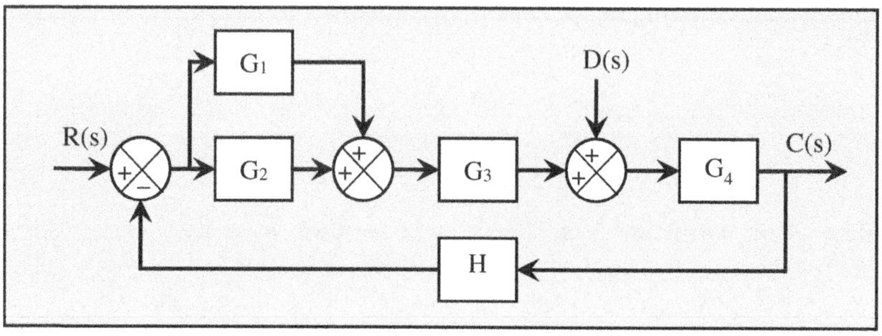

Figure 3.9 The original block diagram of Example 3.9

$$G_1 = \frac{1}{s+10},\ G_2 = \frac{1}{s+1},\ G_3 = \frac{s^2+1}{s^2+4s+4},\ G_4 = \frac{s+1}{s+2},\ H = 2$$

Solution:

Two MATLAB codes are presented here. Both are based on using block diagram reduction rules but in different MATLAB codes.

The First Code:

This code is based on defining the transfer functions as numerators and denominators and showing the plot of the block diagram on the Command Window screen.

1- Enter numerators and denominators of all G's and H blocks as shown in Fig. 3.9.

```
clear
clc
% file name: Ex 3_9_1.m
disp('2 inputs and 1 output system')
nG1=[1]; dG1=[1  10];
nG2=[1]; dG2=[1  1];
nG3=[1  0  1]; dG3=[1  4  4];
nG4=[1  1]; dG4=[1  2];
nH=[2]; dH=[1];
```

After Execution, You Will Get:

```
2 inputs and 1 output system
```

2- Combining G_1 and G_2 blocks in parallel and then combining the result in series with G_3 block. The final block diagram for this step is shown in Fig. 3.9-a.

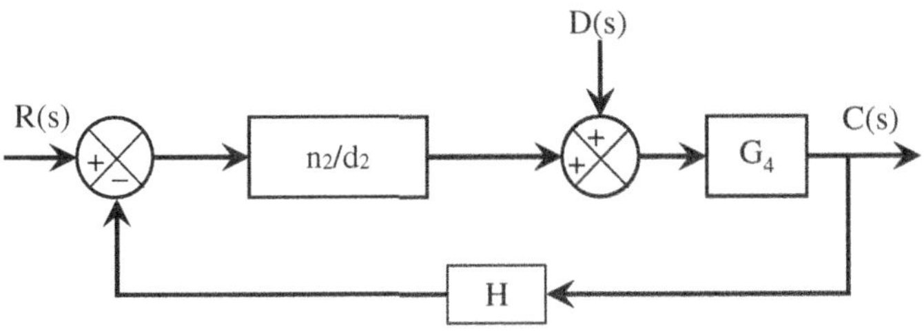

Figure 3.9-a The block diagram after completing step 2

Where $\frac{n_2}{d_2} = (G_1 + G_2)G_3$

```
[n1,d1]=parallel(nG1,dG1,nG2,dG2);
printsys(n1,d1)
[n2,d2]=series(n1,d1,nG3,dG3);
printsys(n2,d2)
```

After Execution, You Will Get:

```
num/den = n1/d1

        2 s + 11
    ----------------
    s^2 + 11 s + 10

num/den = n2/d2

        2 s^3 + 11 s^2 + 2 s + 11
    ----------------------------------------
    s^4 + 15 s^3 + 58 s^2 + 84 s + 40
```

3- To get the C(s)/R(s) transfer function, first, let D(s) = 0, then combine n_2/d_2 and G_4 blocks in series. Next, eliminate the feedback path involving H as shown in Fig. 3.9-b. Notice how we plot the block diagram by using the "`disp`" command in the MATLAB script file.

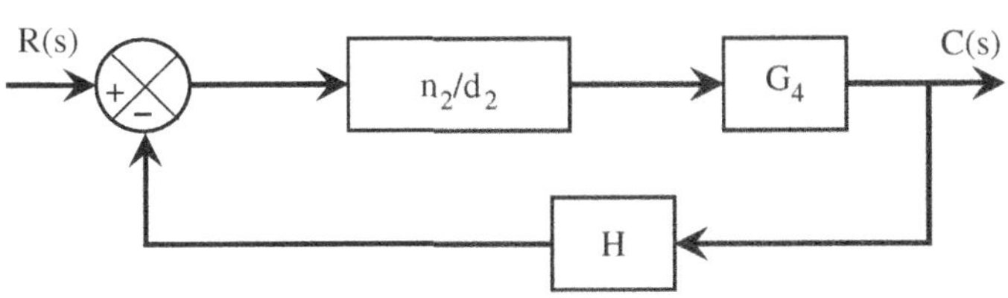

Figure 3.9-b The block diagram after completing step 3.

```
disp('solution to get C(s)/R(s) transfer function')
% plotting the system block diagram
disp('                    _______________      _______          ')
disp('                   |               |    |       |         ')
disp('R------>O---> |       n2/d2       |-->|   G4   |---> C   ')
disp('         -  ^     |_______________|    |_______| |        ')
disp('            |                                     |        ')
disp('            |                                     |        ')
disp('            |                _______              |        ')
disp('            |---------------|   H   |<---------|        ')
disp('                            |_______|                      ')
[n3,d3]=series(n2,d2,nG4,dG4);
printsys(n3,d3)
[n4,d4]=feedback(n3,d3,nH,dH,-1);
disp('C(s)/R(s) transfer function')
printsys(n4,d4)
```

After Execution, You Will Get:

```
solution to get C(s)/R(s) transfer function
                    _______________      _______
                   |               |    |       |
R------>O---> |       n2/d2       |-->|   G4   |---> C
         -  ^     |_______________|    |_______| |
            |                                     |
            |                                     |
            |                _______              |
            |               |       |             |
            |---------------|   H   |<---------|
                            |_______|
 num/den =

        2 s^4 + 13 s^3 + 13 s^2 + 13 s + 11
   ----------------------------------------------
   s^5 + 17 s^4 + 88 s^3 + 200 s^2 + 208 s + 80

C(s)/R(s) transfer function
num/den =
          2 s^4 + 13 s^3 + 13 s^2 + 13 s + 11
   ------------------------------------------------
   s^5 + 21 s^4 + 114 s^3 + 226 s^2 + 234 s + 102
```

4- To get the C(s)/D(s) transfer function, first, let R(s) = 0. Second, eliminate the feedback path involving the series combination of n_2/d_2 and H blocks as shown in Fig. 3.9-c. Notice how we plot the block diagram by using the "`disp`" command.

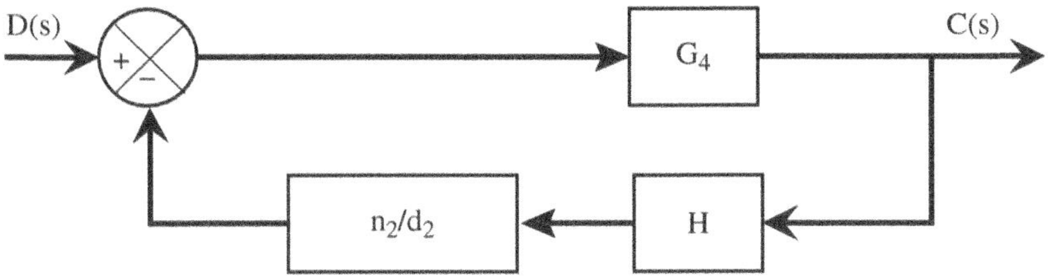

Figure 3.9-c The block diagram after completing step 4.

```
disp('solution to get C(s)/D(s) transfer function')
% plotting the system block diagram
disp('                                  ________                       ')
disp('                                 |        |                      ')
disp('D ---->O----------->  |   G4     |----------------> C ')
disp('        -  ^                     |________|              |        ')
disp('           |                                             |        ')
disp('           |          ____________        ________       |        ')
disp('           |         |            |      |        |      |        ')
disp('           |----|      n2/d2      |<---|    H     |---           ')
disp('                |____________|      |________|                  ')
 [n5,d5]=series(n2,d2,nH,dH);
printsys(n5,d5)
 [n6,d6]=feedback(nG4,dG4,n5,d5,-1);
disp('C(s)/D(s) transfer function')
printsys(n6,d6)
```

After Execution, You Will Get:

```
solution to get C(s)/D(s) transfer function

                              ________
                             |        |
 D ---->O---------->  |   G4     |----------------> C
      -  ^                   |________|              |
         |                                           |
         |                                           |
         |          ____________        ________     |
         |         |            |      |        |    |
         |----|      n2/d2      |<---|    H     |---
              |____________|      |________|
```

```
num/den =

        4 s^3 + 22 s^2 + 4 s + 22
   ---------------------------------
   s^4 + 15 s^3 + 58 s^2 + 84 s + 40

C(s)/D(s) transfer function

num/den =

   s^5 + 16 s^4 + 73 s^3 + 142 s^2 + 124 s + 40
   ---------------------------------------------
   s^5 + 21 s^4 + 114 s^3 + 226 s^2 + 234 s + 102
```

you can write the program differently by using the " tf " MATLAB command rather than the numerator and denominator form as shown in the second code.

The Second Code:

This code is using the "tf" command in defining the transfer function.

```
clear
clc
% Another programming procedure Ex 3.9
% file name EX 3_9_2.m
s=tf('s');
G1=1/(s+10);
G2= 1/(s+1);
G3=(s^2+1)/(s^2+4*s+4);
G4=(s+1)/(s+2);
H=2;
sys1=parallel(G1,G2)
sys2=series(sys1,G3)
disp('solution to get C(s)/R(s) transfer function')
sys3=series(sys2,G4)
disp('C(s)/R(s) transfer function')
sys4=feedback(sys3,H,-1)
disp('solution to get C(s)/D(s) transfer function')
sys5=series(sys2,H)
disp('C(s)/D(s) transfer function')
sys6=feedback(G4,sys5,-1)
```

After Execution, You Will Get The Same Results as Those of Program 1:

```
sys1 =

      2 s + 11
  ----------------
  s^2 + 11 s + 10

sys2 =

      2 s^3 + 11 s^2 + 2 s + 11
  -----------------------------------
  s^4 + 15 s^3 + 58 s^2 + 84 s + 40

solution to get C(s)/R(s) transfer function

sys3 =

      2 s^4 + 13 s^3 + 13 s^2 + 13 s + 11
  ----------------------------------------------
  s^5 + 17 s^4 + 88 s^3 + 200 s^2 + 208 s + 80

C(s)/R(s) transfer function

sys4 =

       2 s^4 + 13 s^3 + 13 s^2 + 13 s + 11
  ------------------------------------------------
  s^5 + 21 s^4 + 114 s^3 + 226 s^2 + 234 s + 102

solution to get C(s)/D(s) transfer function

sys5 =

      4 s^3 + 22 s^2 + 4 s + 22
  ---------------------------------
  s^4 + 15 s^3 + 58 s^2 + 84 s + 40

C(s)/D(s) transfer function

sys6 =

    s^5 + 16 s^4 + 73 s^3 + 142 s^2 + 124 s + 40
  ------------------------------------------------
  s^5 + 21 s^4 + 114 s^3 + 226 s^2 + 234 s + 102
```

For More Practice:

See example 2-4 on page 48 [3], knowing that this example in Ogata was solved in a different way.

Example 3.10 Obtain the $\frac{C_1(s)}{R_1(s)}, \frac{C_1(s)}{R_2(s)}, \frac{C_2(s)}{R_1(s)}, \frac{C_2(s)}{R_2(s)}$ transfer functions for the two-input, two-output system of Fig. 3.10. This example is closer to Ogata [3, p: 49, Ex. 2-5].

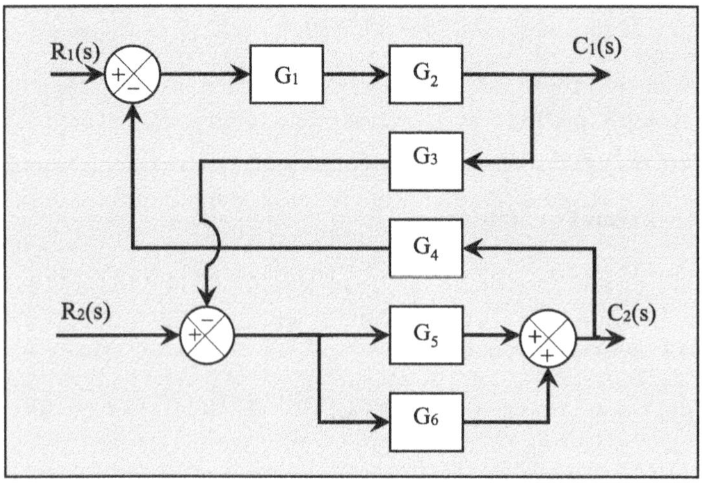

Figure 3.10 The original block diagram of Example 3.10

$$G_1 = \frac{1}{s+10}, G_2 = \frac{1}{s+1}, G_3 = \frac{s^2+1}{s^2+4}, G_4 = \frac{s+1}{s+2}, G_5 = 2, G_6 = \frac{1}{s+2}$$

Solution:

1- Start with "`clear`" and "`clc`" then enter all G's transfer functions.

```
clear
clc
% file name: Ex 3_10.m
s=tf('s');
G1=1/(s+10);
G2= 1/(s+1);
G3=(s^2+1)/(s^2+4);
G4=(s+1)/(s+2);
G5=2;
G6=1/(s+2);
```

2- G_1 and G_2 blocks are combined in series, while G_5 and G_6 blocks are combined in parallel. The results are shown in Fig. 3.10-a.

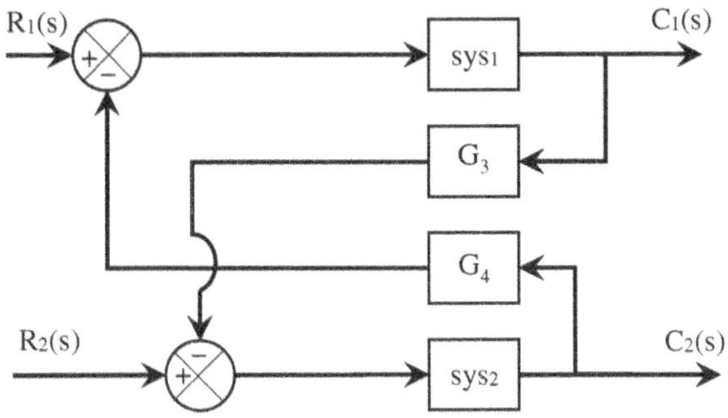

Figure 3.10-a The block diagram after completing step 2.

```
11- sys1=series(G1,G2)
12- sys2=parallel(G5,G6)
```

After Execution:

```
sys1 =
           1
  ---------------
  s^2 + 11 s + 10

sys2 =

  2 s + 5
  -------
   s + 2
```

3- To get $C_1(s)/R_1(s)$: first, combining $-G_3$, sys_2, and G_4 blocks in series gives sys_4 as shown in Figs. 3.10-b and 3.10-c. Second, eliminating the feedback involving $-sys_4$ gives sys_5 as shown in Fig. 3.10-d.

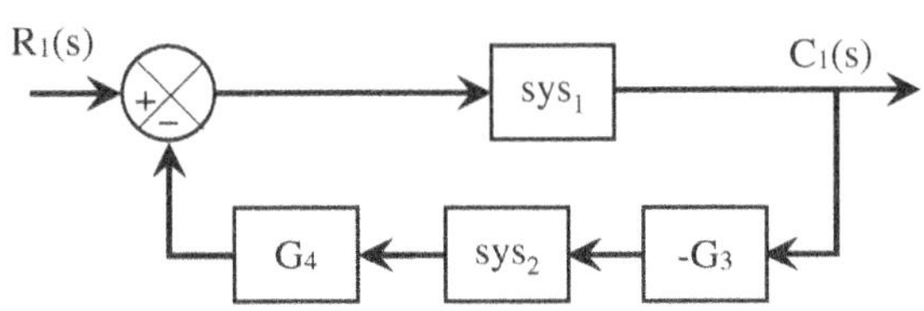

Figure 3.10-b The block diagram of step 3.

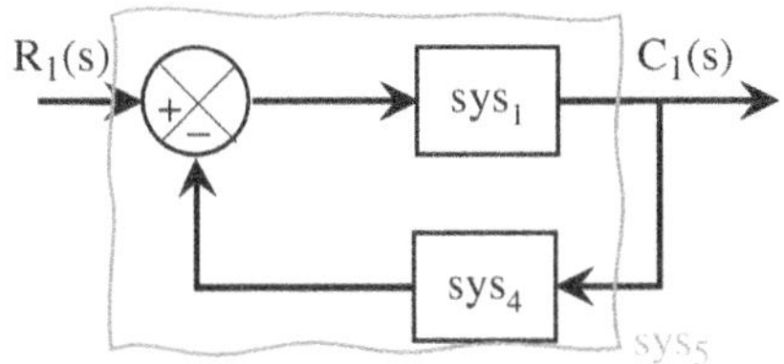

Figure 3.10-c Reduced block diagram of step 3.

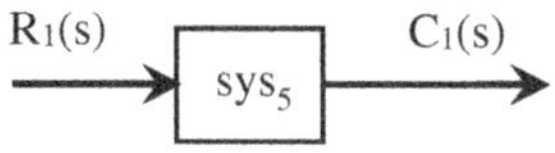

Figure 3.10-d The block diagram after completing step 3.

```
disp('solution to get c1(s)/R1(s) transfer function')
sys3=series(sys2,-1*G3);
sys4=series(sys3,G4)
disp('C1(s)/R1(s) transfer function')
sys5=feedback(sys1,sys4,-1)
```

After Execution:

```
solution to get C1(s)/R1(s) transfer function

sys4 =

  -2 s^4 - 7 s^3 - 7 s^2 - 7 s - 5
  ----------------------------------
  s^4 + 4 s^3 + 8 s^2 + 16 s + 16
```

```
C1(s)/R1(s) transfer function
sys5 =
                  s^4 + 4 s^3 + 8 s^2 + 16 s + 16
  ------------------------------------------------------
  s^6 + 15 s^5 + 60 s^4 + 137 s^3 + 265 s^2 + 329 s + 155
```

4- To get $C_1(s)/R_2(s)$: first, combine $-G_4$, sys_2, and sys_1 blocks in series then eliminate the feedback involving $-G_3$.

```
disp('solution to get C1(s)/R2(s) transfer function')
sys6=series(sys2,-1*G4);
sys7=series(sys6,sys1)
disp('C1(s)/R2(s) transfer function')
sys8=feedback(sys7,G3,-1)
```

After Execution:

```
solution to get C1(s)/R2(s) transfer function
sys7 =
          -2 s^2 - 7 s - 5
  ----------------------------------
  s^4 + 15 s^3 + 58 s^2 + 84 s + 40
```

```
C1(s)/R2(s) transfer function
sys8 =

            -2 s^4 - 7 s^3 - 13 s^2 - 28 s - 20
  ------------------------------------------------------
  s^6 + 15 s^5 + 60 s^4 + 137 s^3 + 265 s^2 + 329 s + 155
```

Activity:

Draw the block diagram for this case.

5- To get $C_2(s)/R_1(s)$: first, combining sys_1, $-G_3$ and sys_2 blocks in series gives sys_{10} as shown in Figs. 3.10-e and 3.10-f. Second, eliminating the feedback involving $-G_4$ gives sys_{11} as shown in Fig. 3.10-g.

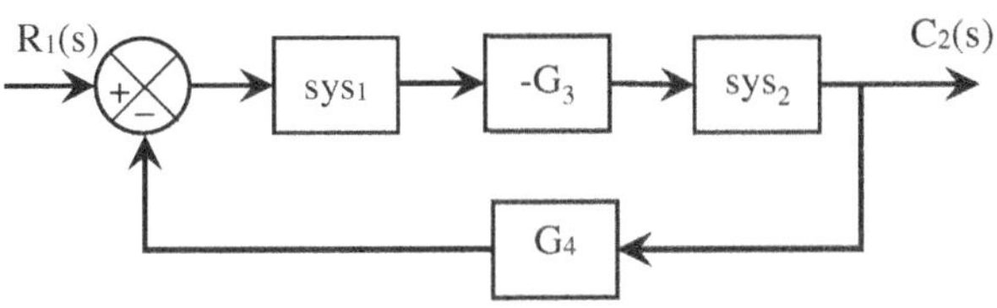

Figure 3.10-e The block diagram of step 5.

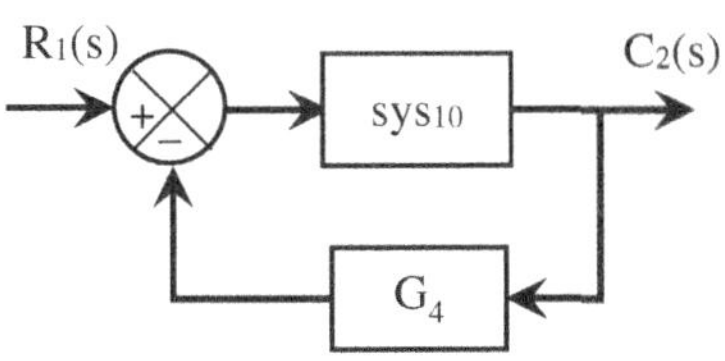

Figure 3.10-f Reduced block diagram of step 5.

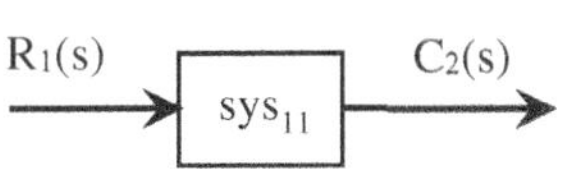

Figure 3.10-g The block diagram after completing step 5.

```
disp('solution to get C2(s)/R1(s) transfer function')
sys9=series(sys1,-1*G3);
sys10=series(sys9,sys2)
disp('C2(s)/R1(s) transfer function')
sys11=feedback(sys10,G4,-1)
```

After Execution:

```
solution to get C2(s)/R1(s) transfer function
sys10 =
             -2 s^3 - 5 s^2 - 2 s - 5
  ------------------------------------------
  s^5 + 13 s^4 + 36 s^3 + 72 s^2 + 128 s + 80
```

```
C2(s)/R1(s) transfer function
sys11 =

             -2 s^4 - 9 s^3 - 12 s^2 - 9 s - 10
  -------------------------------------------------------
  s^6 + 15 s^5 + 60 s^4 + 137 s^3 + 265 s^2 + 329 s + 155
```

6- To get $C_2(s)/R_2(s)$: combine sys_1, $-G_4$ and G_3 blocks in series then eliminate the feedback involving $-sys_{13}$.

```
disp('solution to get C2(s)/R2(s) transfer function')
sys12=series(sys1,-1*G4);
sys13=series(sys12,G3)
disp('C2(s)/R2(s) transfer function')
sys14=feedback(sys2,sys13,-1)
```

After Execution:

```
solution to get c2(s)/R2(s) transfer function

sys13 =
              -s^3 - s^2 - s - 1
  ------------------------------------------
  s^5 + 13 s^4 + 36 s^3 + 72 s^2 + 128 s + 80
```

```
C2(s)/R2(s) transfer function
```

```
sys14 =

  2 s^6 + 31 s^5 + 137 s^4 + 324 s^3 + 616 s^2 + 800s + 400
  ------------------------------------------------------------
     s^6 + 15 s^5 + 60 s^4 + 137 s^3 + 265 s^2 + 329 s + 155
```

Activity:

Draw the block diagram for this case.

For More Practice:

See example A-2-5 on page 48 [3], which is closer to our example. The difference is in the solving procedure.

Example 3.11 Obtain the overall transfer function for the system of Fig. 3.11 [3, p: 60, Problem B-2-2].

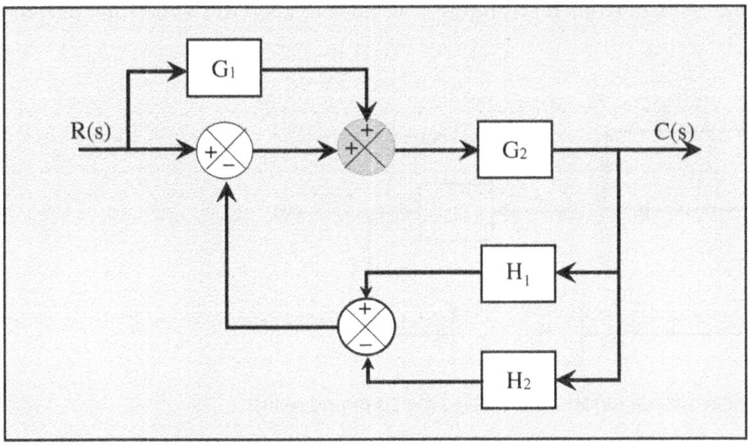

Figure 3.11 The original block diagram of Example 3.11

$$G_1 = \frac{1}{s+10},\ G_2 = \frac{1}{s+1},\ H_1 = \frac{s+1}{s+2}, H_2 = 2$$

Solution:

Two methods are presented to solve this example; the first one is based on using the rules of block diagram reduction, while the second one is based on writing the outputs after each block in equation form and then solving them.

The First Method:

1- Rearrange the first and second summing points then combine H_1 and H_2 blocks in parallel as illustrated in Fig. 3.11-a.

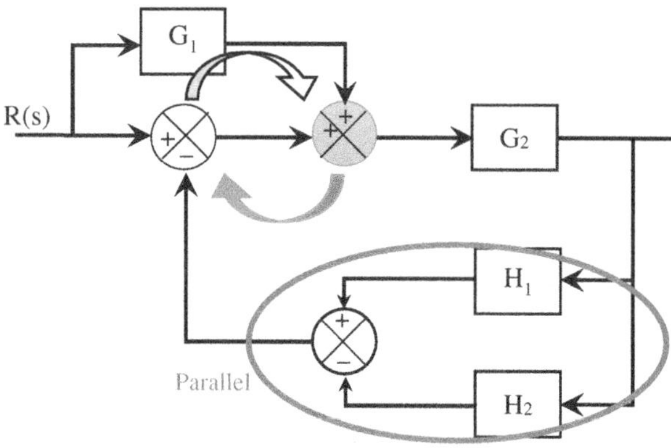

Figure 3.11-a The block diagram after completing step 1.

Before rearranging the summing points, you have to know all the inputs and outputs for each summing point as shown in Fig. 3.11-b. After rearranging, you have to keep the same final output as shown in Fig. 3.11-c.

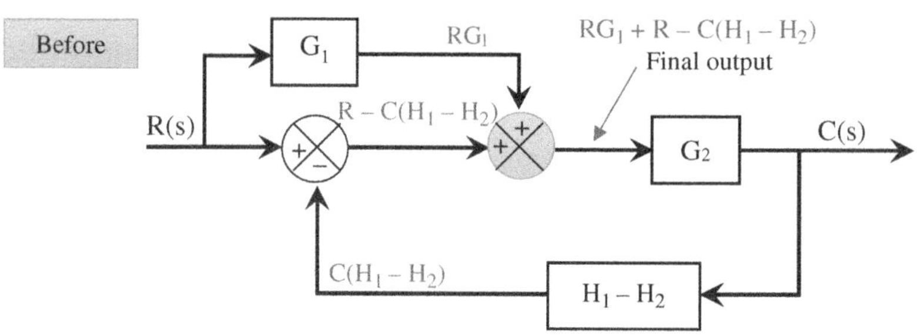

Figure 3.11-b The output after each block before rearranging the summing points.

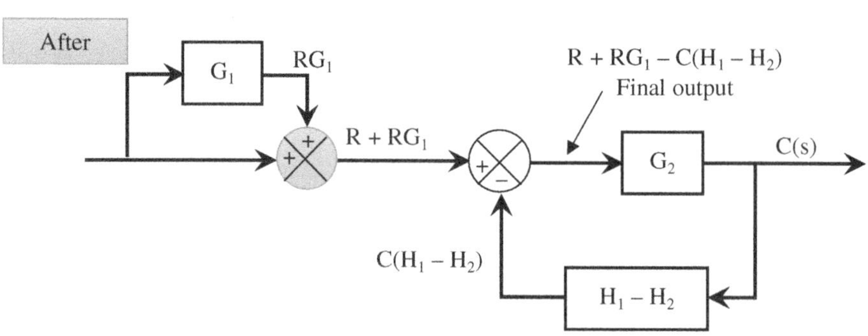

Figure 3.11-c The output after each block after rearranging the summing points.

2- Combining block G_1 with 1 in parallel and eliminating the feedback loop formed by H_1-H_2 gives Fig. 3.11-d.

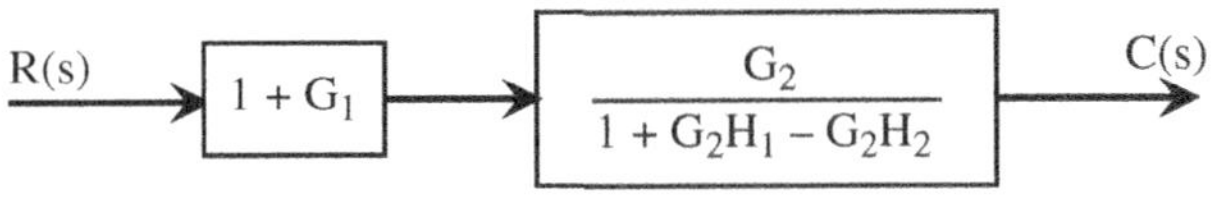

Figure 3.11-d The block diagram after completing step 2.

3- After combining the two blocks of Fig. 3.11-d in series, the final transfer function will be:

$$\frac{C(s)}{R(s)} = \frac{G_2(1 + G_1)}{1 + G_2H_1 - G_2H_2}$$

The MATLAB program will be:

```
clear
clc
% file name Ex 3_11.m
s=tf('s');
G1=1/(s+10);
G2=1/(s+1);
H1=(s+1)/(s+2); H2=2;
disp('Combining H1 and H2 blocks in parallel gives H') % display enclosed text
H=parallel(H1,-H2)
disp('Combining G1 block and 1 in parallel gives G3') % display enclosed text
G3=parallel(G1,1)
disp('Elimination of the feedback loop formed by H1-H2 block gives G4') % display enclosed text
G4=feedback(G2,H,-1)
disp('The final transfer function C(s)/R(s)will be G5') % display enclosed text
G5=series(G3,G4)
```

After Execution, The Results Will Be:

```
'Combining H1 and H2 blocks in parallel gives H'
H =
  -s - 3
  ------
   s + 2
```

```
'Combining G1 block and 1 in parallel gives G3'
G3 =
  s + 11
  ------
  s + 10

'Elimination of the feedback loop formed by H1-H2 block gives G4'
G4 =
       s + 2
  -------------
  s^2 + 2 s - 1
'The final transfer function C(s)/R(s)will be G5'

G5 =
       s^2 + 13 s + 22
  ------------------------
  s^3 + 12 s^2 + 19 s - 10
```

For More Practice:

See also problem B-2-1 on page 60 [3].

The Second Method:

In this method, the output will be obtained after each block in equation form until reaching the final output C as shown in Fig. 3.11-e. Write an equation for E, then write another one for C as shown in Eqs. (3.1) and (3.2). Next, simplify the equation until you get C(s)/R(s).

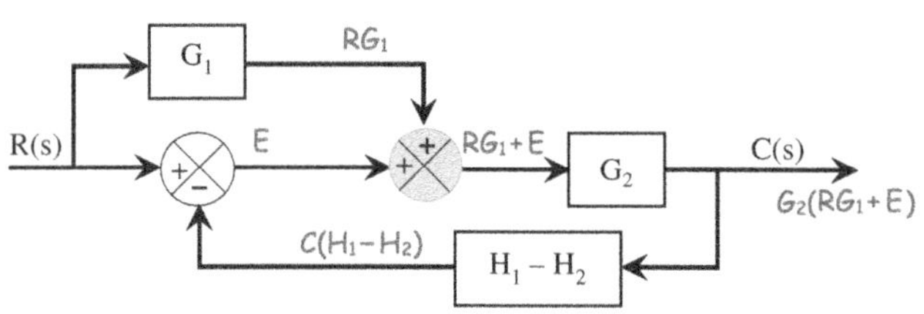

Figure 3.11-e The output after each block.

$$\text{The error equation: } E = R - C(H_1 - H_2) \tag{3.1}$$

$$\text{The output equation: } C = G_2(RG_1 + E) \tag{3.2}$$

Substitute Eq. (3.1) into Eq. (3.2) to get:

$$C = G_2(RG_1 + R - C(H_1 - H_2)) \tag{3.2$'$}$$

After simplification, we will get:

$$\frac{C(s)}{R(s)} = \frac{G_2(1+G_1)}{1+G_2H_1 - G_2H_2}$$

Note:

- Write the output of the first summing point as $R - C(H_1 - H_2)$ directly without E because the block diagram of Example 3.11 is simple, and then use Eq. (3.2′) to get C(s)/R(s).
- In case of a bigger block diagram, I recommend using E for a simpler process for getting C(s)/R(s).

Example 3.12 Find the C(s)/R(s) for the system of Fig. 3.12 [2, p: 120, problem 15-a]

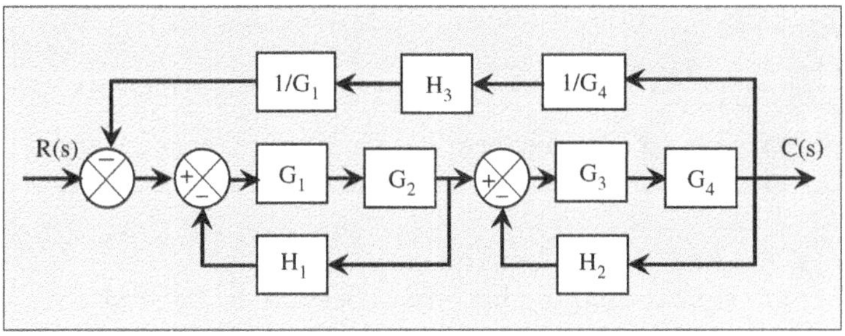

Figure 3.12 The original block diagram of Example 3.12

$$G_1 = \frac{1}{s+10},\quad G_2 = \frac{1}{s+1},\quad G_3 = \frac{s^2+1}{s^2+4s+4},\quad G_4 = \frac{s+1}{s+6},$$

$$H_1 = \frac{s+1}{s+2},\quad H_2 = 2,\quad H_3 = 1$$

Solution:

1- Combining G_1 and G_2 blocks in series, then eliminating the negative feedback loop formed by H_1 gives G_6.

2- Combining G_3 and G_4 blocks in series, then eliminating the negative feedback loop formed by H_2 gives G_8.

3- Combining $1/G_1$, H_3, and $1/G_4$ blocks in series gives G_{10}.

4- After doing steps 1, 2, and 3, the block diagram will be as shown in Fig. 3.12-a.

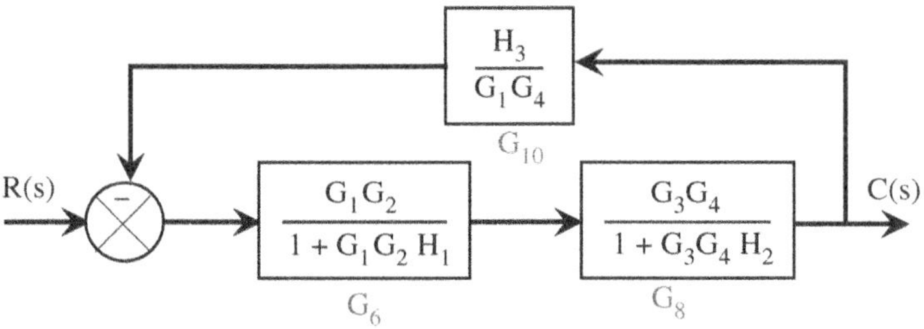

Figure 3.12-a The block diagram after completing steps 1–4

5- Combining G_6 and G_8 blocks in series and eliminating the feedback loop formed by block G_{10} gives the final transfer function.

$$\frac{C(s)}{R(s)} = \frac{G_1G_2G_3G_4}{1 + G_1G_2H_1 + G_2G_3H_3 + G_3G_4H_2 + G_1G_2G_3G_4H_1H_2}$$

The MATLAB code will be:

```
clear
clc
% file name: Ex 3_12.m
s=tf('s');
% Define all G's and H's blocks
G1=1/(s+10); G2=1/(s+1);
G3=(s^2+1)/(s^2+4*s+1); G4=(s+1)/(s+6);
H1=(s+1)/(s+2); H2=2; H3=1;
G5=series(G1,G2); % G1 and G2 in series
G6=feedback(G5,H1,-1) % Eliminate the -ve. feedback formed by H1
G7=series(G3,G4); % G3 and G4 in series
G8=feedback(G7,H2,-1) % Eliminate the -ve. feedback formed by H2
G9=series(1/G1,H3); %%% 1/G1 and H3 in series
G10=series(G9,1/G4) %%% G9 and 1/G4 in series
G11=series(G6,G8); % G6 and G8 in series
G12=feedback(G11,G10,-1)% Eliminate the -ve. feedback formed by G10
```

After Execution:

```
G6 =

             s + 2
  ----------------------
  s^3 + 13 s^2 + 33 s + 21

G8 =

       s^3 + s^2 + s + 1
  -------------------------
  3 s^3 + 12 s^2 + 27 s + 8

G10 =
       s^2 + 16 s + 60
       ---------------
           s + 1
```

```
G12 =
          s^5 + 4 s^4 + 6 s^3 + 6 s^2 + 5 s + 2
------------------------------------------------------------
  3 s^7 + 55 s^6 + 352 s^5 + 1211 s^4 + 2296 s^3 + 2308 s^2 +
1211s +288
```

Example 3.13 Find the transfer function for the system of Fig. 3.13 [2, p: 120, problem 15-b].

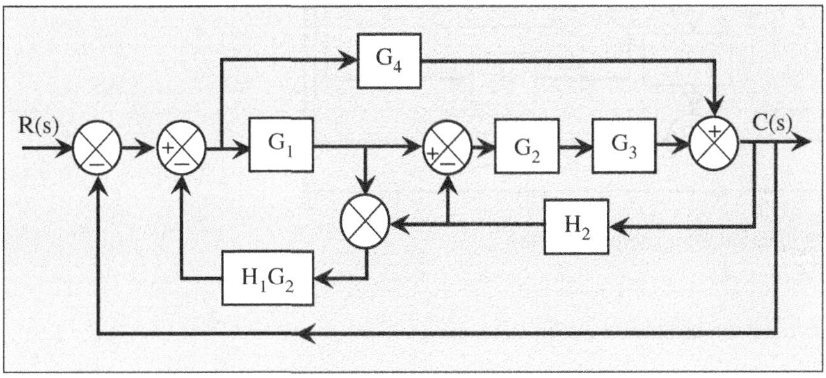

Figure 3.13 The original block diagram of Example 3.13

Solution:

Two methods are presented to solve the block diagram of Fig. 3.13; the first one is based on using the rules of block diagram reduction, and the second one is based on equations.

The First Method:

1- Figure 3.13 can be redrawn as shown in Fig. 3.13-a.

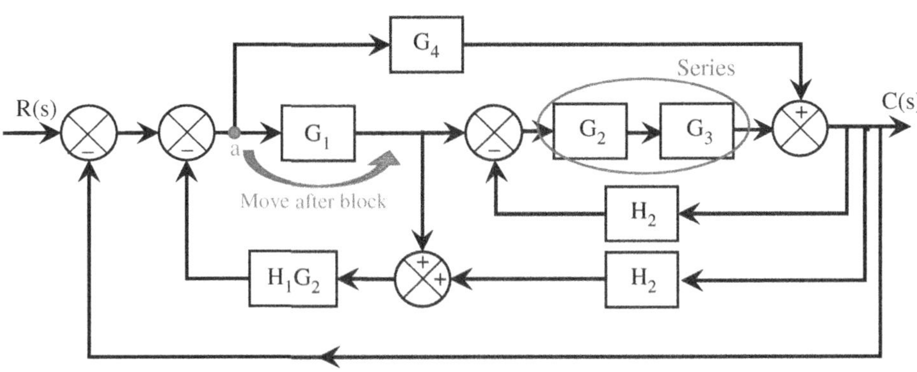

Figure 3.13-a Another way to draw the block diagram of Fig. 3.13.

2- * Moving the takeoff point "a" after block G_1 as illustrated in Fig. 3.13-a gives G_4/G_1.

* Combining G_2 and G_3 in series as illustrated in Fig. 3.13-a gives G_2G_3.
* Changing the place of block H_1G_2. The results will be as shown in Fig. 3.13-b.

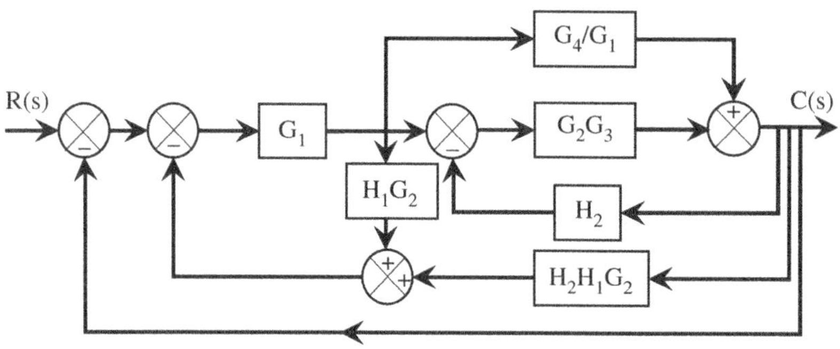

Figure 3.13-b The block diagram after completing step 2.

3- Figure 3.13-b can be redrawn as shown in Fig. 3.13-c.

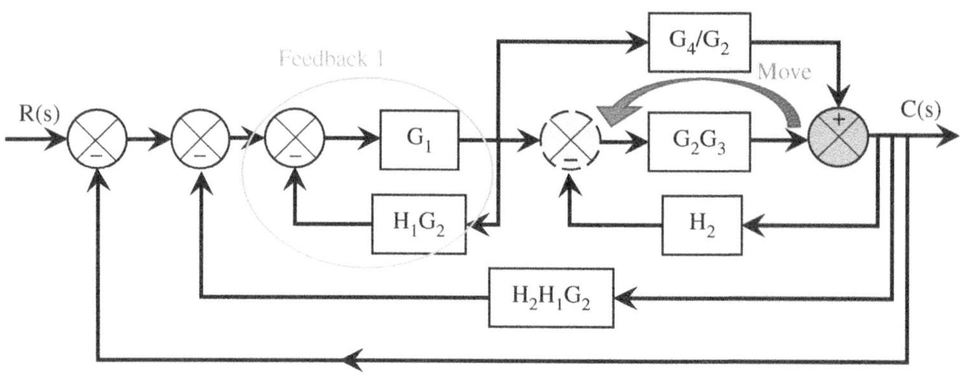

Figure 3.13-c Another way to draw the block diagram of Fig. 3.13-b.

4- Eliminating the negative feedback loop formed by block H_1G_2 as illustrated in Fig. 3.13-c as Feedback 1 gives $G_5 = G_1/(1 + H_1G_2G_1)$. Then moving the last summing point before block G_2G_3 gives

$$G_6 = \frac{\frac{G_4}{G_1}}{G_2G_3} = \frac{G_4}{G_1G_2G_3},$$ as shown in Fig. 3.13-d.

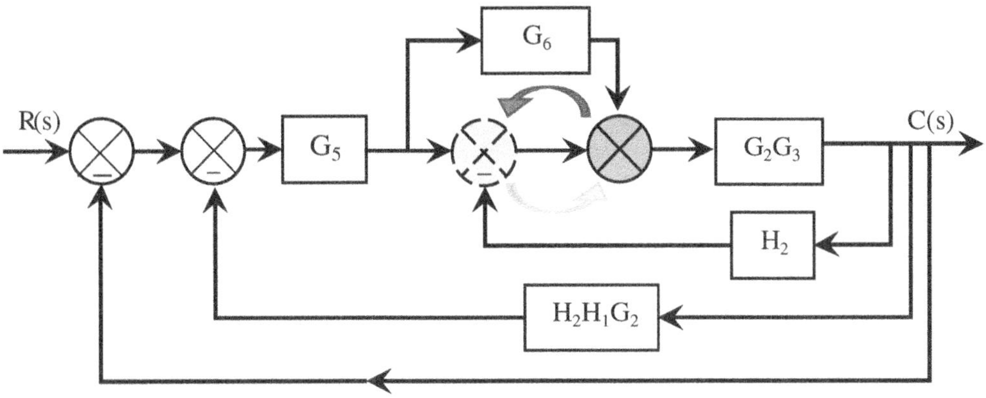

Figure 3.13-d The block diagram after completing step 4

Where $G_5 = \dfrac{G_1}{1 + G_1G_2H_1}$, $G_6 = \dfrac{G_4/G_1}{G_2G_3} = \dfrac{G_4}{G_1G_2G_3}$

5- Interchanging the 3rd and 4th summing points as demonstrated in Fig. 3.13-d gives Fig. 3.13-e.

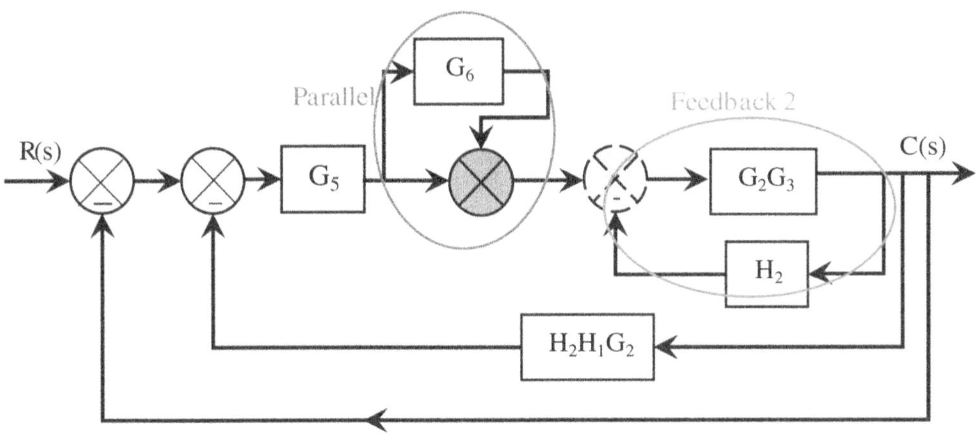

Figure 3.13-e The block diagram after completing step 5.

6- * Eliminating the feedback loop formed by H_2 gives $G_7 = \frac{G_2G_3}{1+G_2G_3H_2}$.
* Combining the two parallel paths G_6 and 1 gives $G_8 = G_6 + 1$.

See Figs. 3.13-e and 3.13-f.

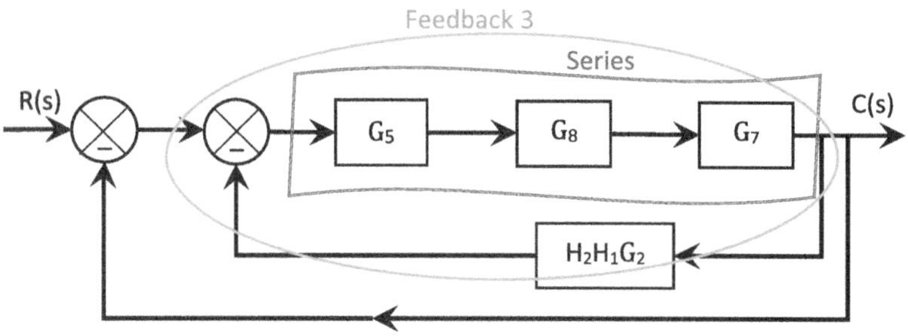

Figure 3.13-f The block diagram after completing step 6

Where $G_7 = \dfrac{G_2G_3}{1+G_2G_3H_2}$, $G_8 = G_6 + 1 = \dfrac{G_4 + G_1G_2G_3}{G_1G_2G_3}$

7- * Combing the three blocks G_5, G_8, and G_7 in series gives $G_9 = G_5G_8G_7$ as shown in Fig. 3.13-f.
 * Eliminating the feedback loop formed by $H_2H_1G_2$ gives G_{10} as shown in Figs. 3.13-f and 3.13-g.

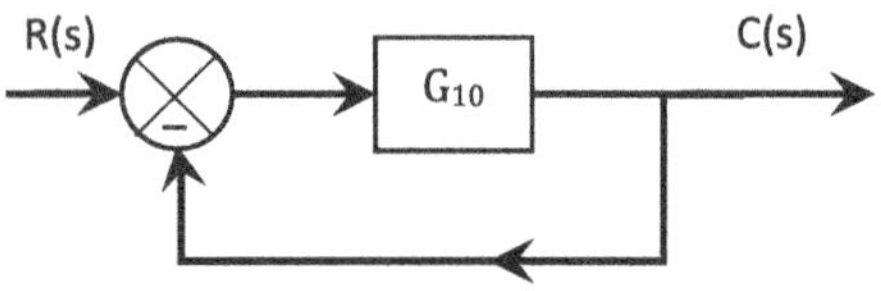

Figure 3.13-g The block diagram after completing step 7

Where $G_9 = G_5G_8G_7 = \dfrac{G_4 + G_1G_2G_3}{(1+G_2G_3H_2)(1+G_1G_2H_1)}$

$$G_{10} = \frac{G_9}{1+G_9H_1H_2G_2} = \frac{G_4 + G_1G_2G_3}{(1+G_2G_3H_2)(1+G_1G_2H_1) + (G_4 + G_1G_2G_3)(H_2H_1G_2)}$$

8- Eliminate the closed loop of Fig. 3.13-g gives G_{11}, which is the final transfer function as shown in Fig. 3.13-h.

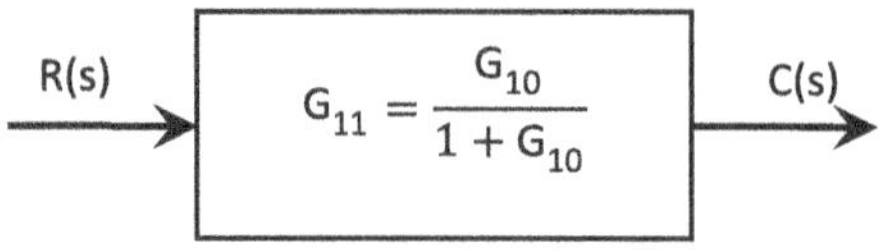

Figure 3.13-h The block diagram after completing step 8.

$$\frac{C(s)}{R(s)} = G_{11} = \frac{G_4 + G_1G_2G_3}{(1+G_2G_3H_2)(1+G_1G_2H_1) + (1+H_2H_1G_2)(G_4 + G_1G_2G_3)}$$

Activity:

- Write a pseudocode for Example 3.13.
- Another idea in solving Example 3.13 is without changing the place of the block H_1G_2 in Fig. 3.13-a:

1- Move the takeoff point "a" in Fig. 3.13-a after block G_1.
2- Move the last summing point before block G_2G_3.
3- Interchange the 3rd and 4th summing points.
4- Eliminate the negative feedback formed by H_2.
5- Combine the parallel paths 1 and $\frac{G_4/G_1}{G_2G_3}$.
6- Combine blocks of steps 4 and 5 in series.
7- Move the takeoff point off "b" after the block of step 4.
8- After finishing the steps from 1 to 7, you will get Fig. 3.13-i.
9- Continue until getting C(s)/R(s).

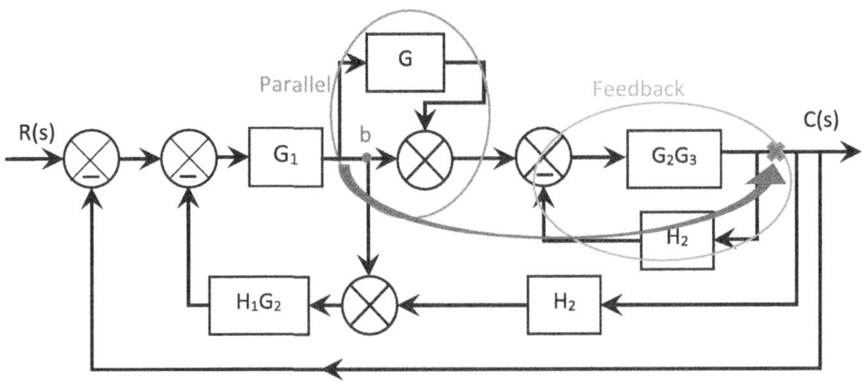

Figure 3.13-i The block diagram after completing steps 1–7

Where $G = \dfrac{G_4/G_1}{G_2G_3}$

The Second Method:

Figure 3.13 can be redrawn with the outputs after each as shown in Fig. 3.13-j.

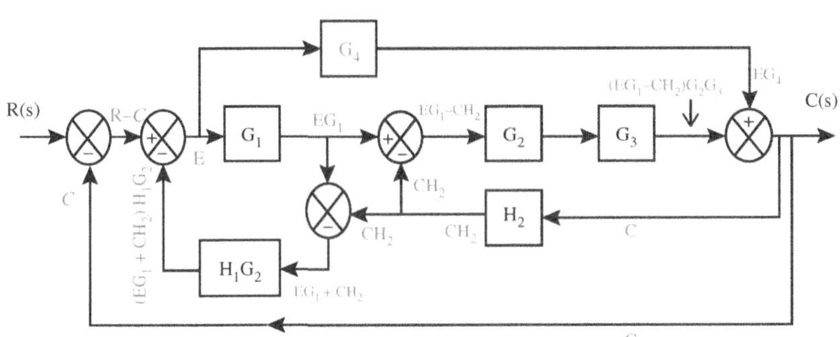

Figure 3.13-j The output after each block.

The first equation will be for E;

$$E = R - C - (EG_1 + CH_2)H_1G_2 \tag{3.3}$$

Simplify Eq. (3.3) until getting an expression for E:

$$E = R - C - EG_1H_1G_2 - CH_2H_1G_2$$

$$E = \frac{R - C(1 + G_2H_1H_2)}{1 + G_1G_2H_1} \tag{3.4}$$

The other equation will be for C;

$$C = EG_4 + (EG_1 - CH_2)G_2G_3 \tag{3.5}$$

Simplify Eq. (3.5) and substitute for E as in Eq. (3.4) until getting C(s)/R(s):

$$C = EG_4 + EG_1G_2G_3 - CH_2G_2G_3$$

$$C + CH_2G_2G_3 = E(G_4 + G_1G_2G_3)$$

$$C(1 + H_2G_2G_3) = E(G_4 + G_1G_2G_3)$$

$$C(1 + H_2G_2G_3) = \left(\frac{R - C(1 + G_2H_1H_2)}{1 + G_1G_2H_1}\right)(G_4 + G_1G_2G_3)$$

$$\boxed{\frac{C(s)}{R(s)} = \frac{G_4 + G_1G_2G_3}{(1 + G_2G_3H_2)(1 + G_1G_2H_1) + (1 + G_2H_1H_2)(G_4 + G_1G_2G_3)}}$$

Example 3.14 Find the transfer function for the block diagram of Fig. 3.14 [2, p: 120, problem 15-c].

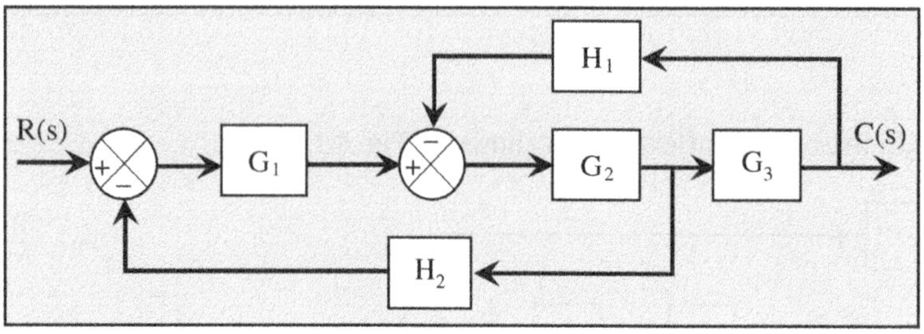

Figure 3.14 The original block diagram of Example 3.14

Solution:

Three methods are illustrated and discussed to reduce the block diagram of Fig. 3.14. The first one focuses on the use of removing a block from a feedback loop rule. The second shows the use of moving a takeoff point behind a block. The third one shows the form of the equation.

The First Method:

1- Moving the takeoff point of the feedback loop formed by H_1 before block G_3 as illustrated in Figs. 3.14-a and 3.14-b.

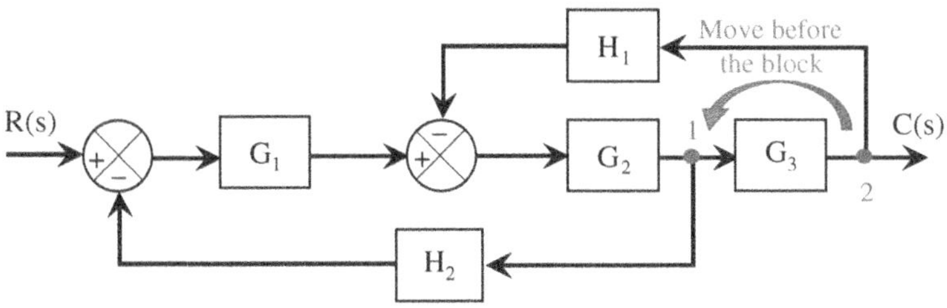

Figure 3.14-a Illustration of a move.

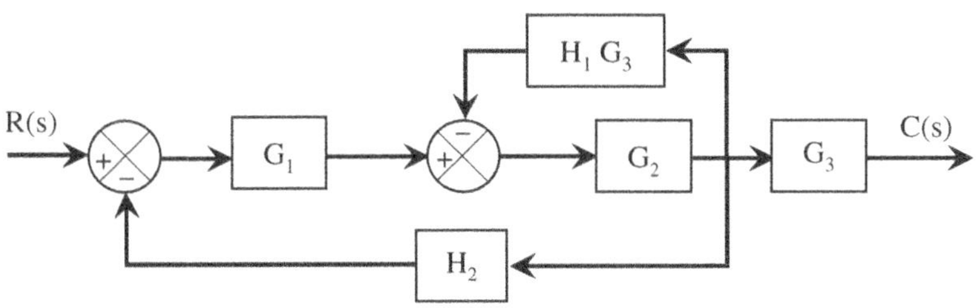

Figure 3.14-b The block diagram after completing step 1.

2- Removing block H_1G_3 from the feedback as shown in Fig. 3.14-c.

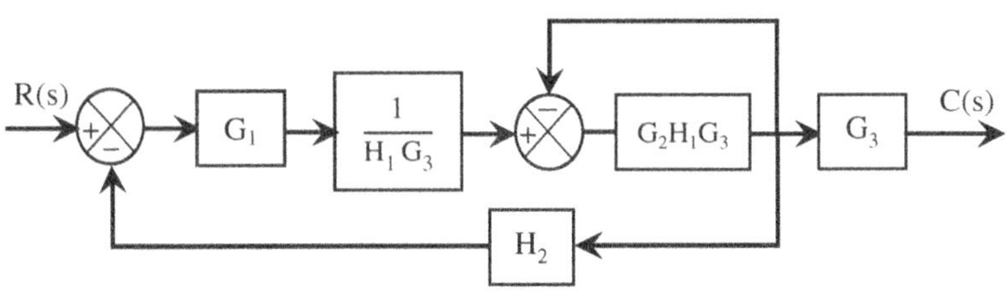

Figure 3.14-c The block diagram after completing step 2.

3- Eliminate the close loop as shown in Fig. 3.14-d.

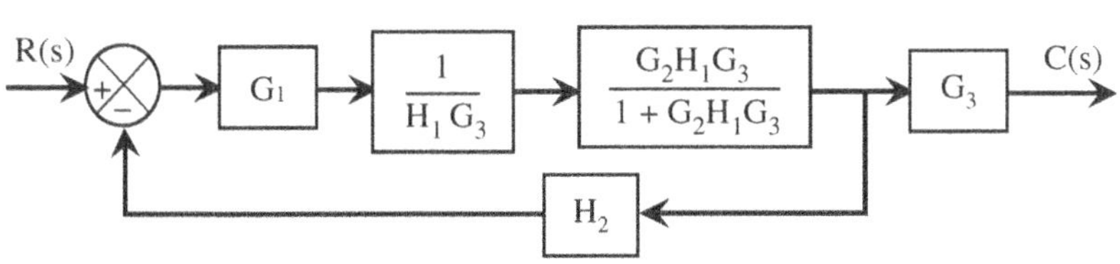

Figure 3.14-d The block diagram after completing step 3.

4- Combine blocks G_1, $\frac{1}{H_1 G_3}$ and $\frac{G_2 H_1 G_3}{1+G_2 H_1 G_3}$ in series as shown in Fig. 3.14-e

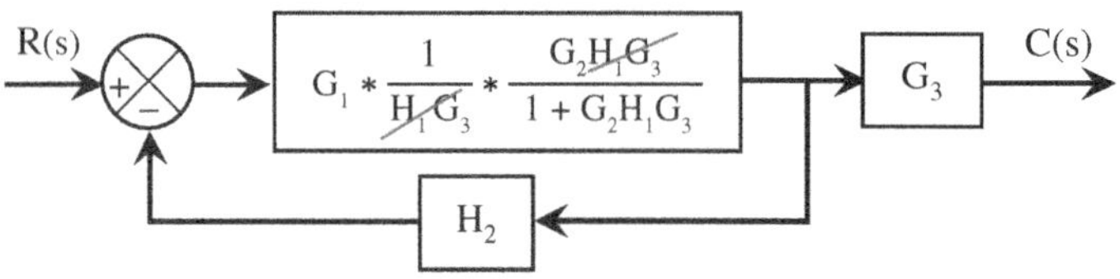

Figure 3.14-e The block diagram after completing step 4.

5- Remove block H_2 from the feedback loop as shown in Fig. 3.14-f.

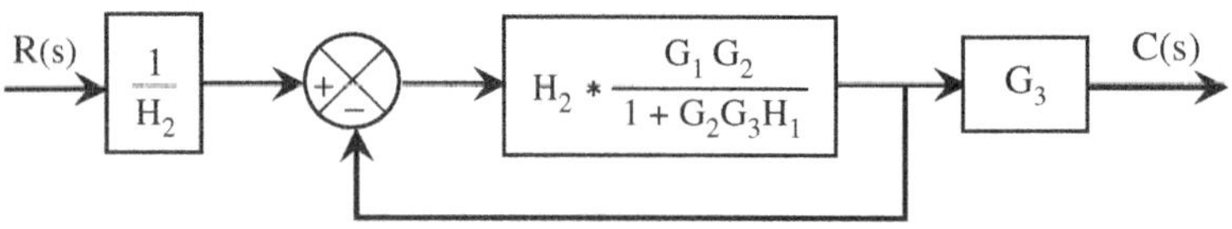

Figure 3.14-f The block diagram after completing step 5.

6- Eliminating the close loop and combining the result with $1/H_2$ and G_3 blocks in series gives Fig. 3.14-g.

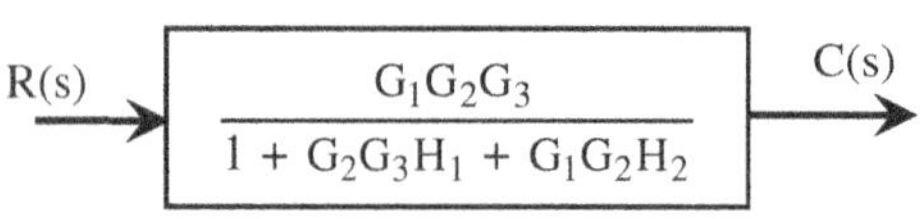

Figure 3.14-g The block diagram after completing step 6.

7- The overall system transfer function will be:

$$\frac{C(s)}{R(s)} = \frac{G_1 G_2 G_3}{1 + G_2 G_3 H_1 + G_1 G_2 H_2}$$

Activity:

Assume transfer functions for all G's and H's blocks, and then write a MATLAB script file. See the following note.

Note:

The block $\frac{1}{H_1 G_3}$ can be represented in Math and MATLAB as follows:

Math:	MATLAB:
$H_1 = \frac{nH_1}{dH_1}$, $G_3 = \frac{nG_3}{dG_3}$, then $\frac{1}{H_1 G_3} = \frac{1}{\frac{nH_1}{dH_1} \cdot \frac{nG_3}{dG_3}} = \frac{dH_1 dG_3}{nH_1 nG_3} = \frac{n_1}{d_1}$	`n1=conv(dH1,dG3),` `d1=conv(nH1,nG3)` `printsys(n1,d1)`

The Second Method:

1- Moving the takeoff point of block H_2 after block G_3 as illustrated in Fig. 3.14-h gives Fig. 3.14-i.

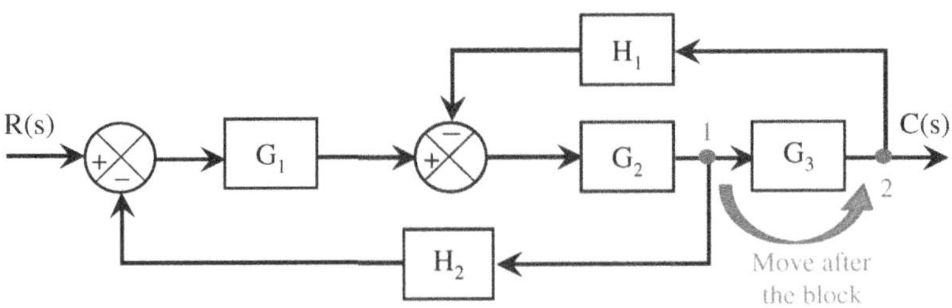

Figure 3.14-h Illustration of a move

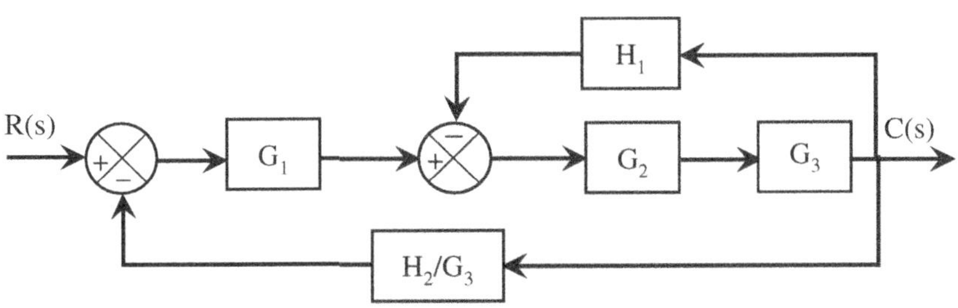

Figure 3.14-i The block diagram after completing step 1

2- Combine G_2 and G_3 blocks in series, then eliminate the feedback loop formed by block H_1 as shown in Fig. 3.14-j.

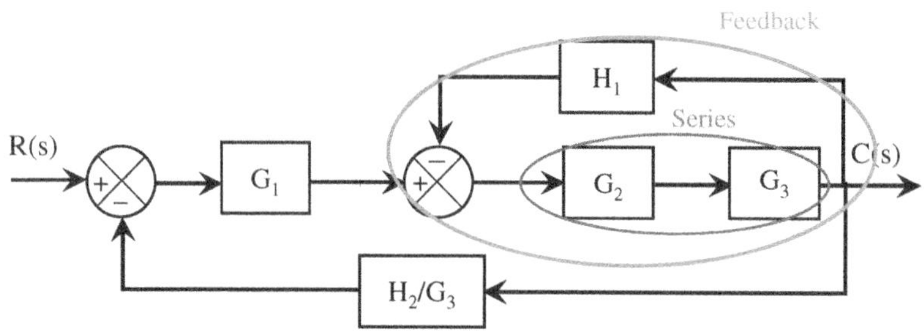

Figure 3.14-j Illustration of step 2

3- The result of step 2 is combined in series with block G_1, then the elimination of the feedback loop formed by block H_2/G_3 gives the final transfer function.

$$\frac{C(s)}{R(s)} = \frac{G_1G_2G_3}{1 + G_2G_3H_1 + G_1G_2H_2}$$

which is the same as in the first method.

Activity:

- Draw the block diagram of step 3.
- Write a MATLAB code for the second method.

The Third Method:

In this method, we will write the output after each block in equation form until reaching the output C. Figure 3.14-k shows these outputs. Our goal is to find an equation that contains only C and R. So, to do that first, write an equation for E and another one for C, then simplify these two equations. After that, the substitution of E in equation C gives an equation that has only C and R. Next, make some simplifications then you will get C(s)/R(s).

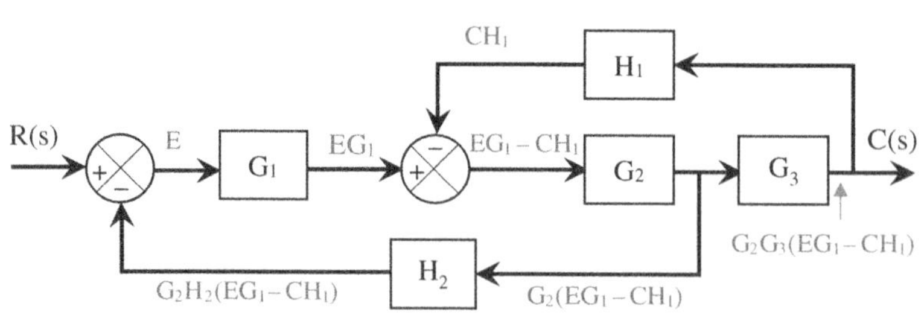

Figure 3.14-k The output after each block.

The error equation: $E = R - G_2H_2(EG_1 - CH_1)$ (3.6)

The output equation: $C = G_2G_3(EG_1 - CH_1)$ (3.7)

After simplifications, Eqs. (3.6) and (3.7) will be:

$$E = \frac{R + CG_2H_1H_2}{1 + G_1G_2H_2} \tag{3.8}$$

$$C(1 + G_2G_3H_1) = EG_1G_2 \tag{3.9}$$

Then substitute Eq. (3.8) in Eq. (3.9) to get:

$$C(1 + G_2G_3H_1) = \left(\frac{R + CG_2H_1H_2}{1 + G_1G_2H_2}\right) G_1G_2G_3 \tag{3.10}$$

After simplification:

$$\frac{C(s)}{R(s)} = \frac{G_1G_2G_3}{1 + G_2G_3H_1 + G_1G_2H_2}$$

Example 3.15 Find the transfer function C(s)/R(s) for the block diagram of Fig. 3.15. This example is closer to problem 15-d on page 120 [2].

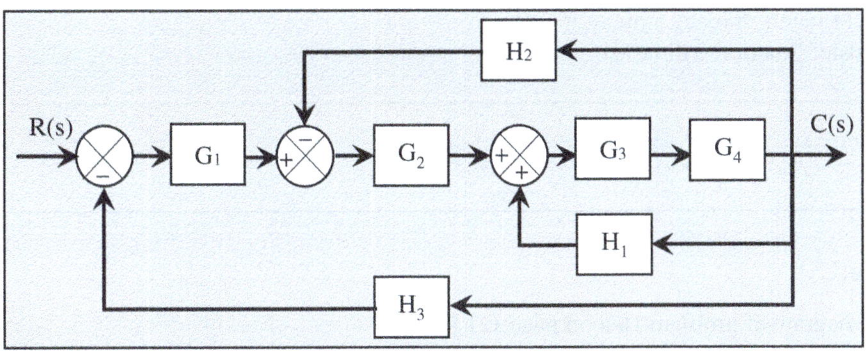

Figure 3.15 The original block diagram of Example 3.15.

Solution:

The pseudocode for this example will be:

1- Combine G_3 and G_4 blocks in series, then eliminate the feedback loop formed by H_1 as shown in Fig. 3.15-a.

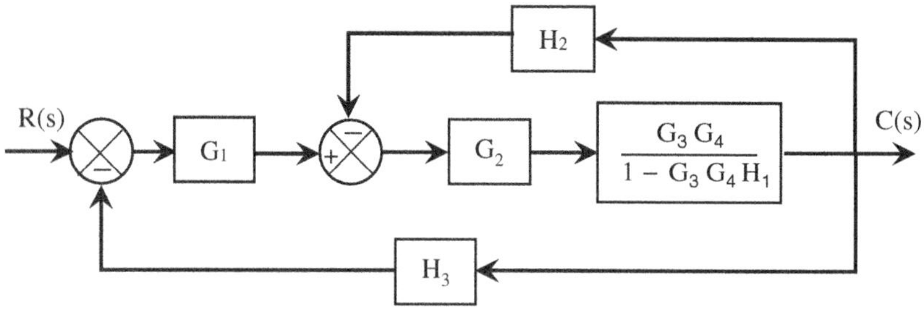

Figure 3.15-a The block diagram after completing step 1.

2- The results of step 1 in series with block G_2.
3- Eliminate the feedback loop formed by H_2.
4- The result of step 3 in series with G_1.
5- Eliminate the feedback loop formed by H_3.

Activity:

- Draw the block diagram for the steps from 2 to 5.
- Write a MATLAB script file for this example, and remember to assume transfer functions for all G's and H's blocks before starting your script file.
  The overall Transfer function will be as:

$$\frac{C(s)}{R(s)} = \frac{G_1G_2G_3G_4}{1 - G_3G_4H_1 + G_2G_3G_4H_2 + G_1G_2G_3G_4H_3}$$

For More Practice:

Reduce the block diagram of problem 16-a on page 121 [2].

Example 3.16 Find the transfer function C(s)/R(s) for the block diagram of Fig. 3.16, knowing that this figure is similar to Fig. 14.17 on page 531 [7] but with different symbols.

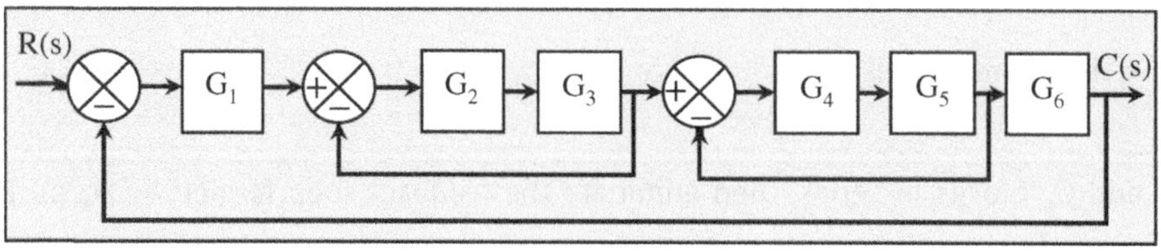

Figure 3.16 The original block diagram of Example 3.16.

Solution:

1- Combine G_2 and G_3 blocks in series and eliminate the closed loop to get G_7.
2- Combine G_4 and G_5 blocks in series and eliminate the closed loop to get G_8. Your block diagram will be as shown in Fig. 3.16-a.

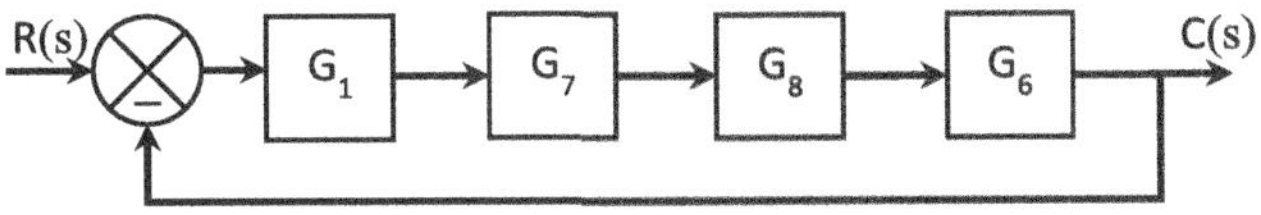

Figure 3.16-a The block diagram after completing steps 1 and 2.

Where $G_7 = G_2G_3$, $G_8 = G_4G_5$

The pseudocode for MATLAB will be:

```
1- Clear
2- clc
3- Name your file.
4- Series(G2 and G3)
5- Cloop (G2G3, 1)
6- Series(G4,G5)
7- Cloop(G4G5,1)
```

3- Combine G_1, G_7, G_8, and G_6 blocks in series and eliminate the closed loop to get C(s)/R(s) as shown in Fig. 3.16-b.

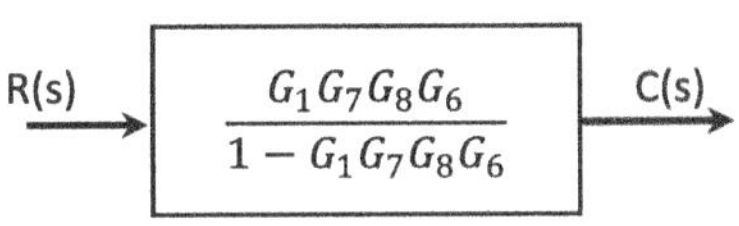

Figure 3.16-b The block diagram after completing step 3.

The pseudocode for MATLAB will be:

```
8- Series the blocks G1,G7, G8 and G6.
9- Cloop (G1G7 G8G6, 1)
```

Activity:

Write the MATLAB code for Example 3.16.

Example 3.17 Obtain C(s)/R(s) for the block diagram of Fig. 3.17 [7, p: 329, Fig. P9.38].

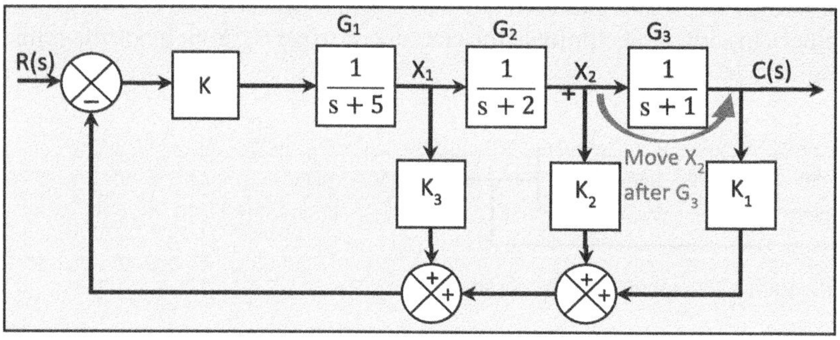

Figure 3.17 The original block diagram of Example 3.17.

Solution:

Two methods are presented to solve this example; the first one is based on using the block diagram reduction rules, while the second one is based on using equations.

The First Method:

1- Combine K and 1/(S + 5) blocks in series.
2- Move the takeoff point 'X_2' after 1/(S + 1) block as illustrated in Fig. 3.17. So, your figure should look like Fig. 3.17-a

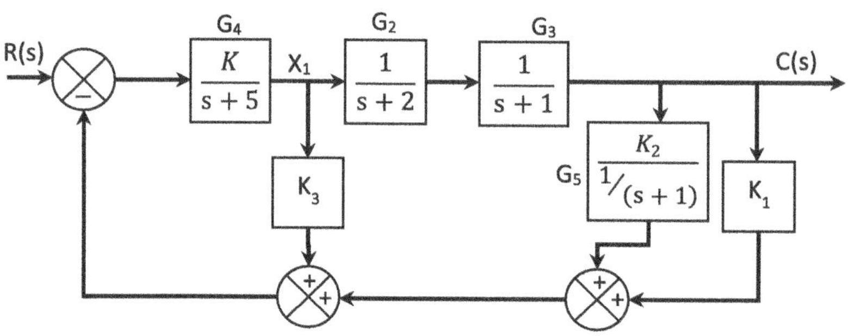

Figure 3.17-a The block diagram after completing step 2.

3- Combining G_2 and G_3 blocks in series gives G_6.
4- Combining G_5 and K_1 blocks in parallel gives G_7. Your block diagram should look like Fig. 3.17-b.

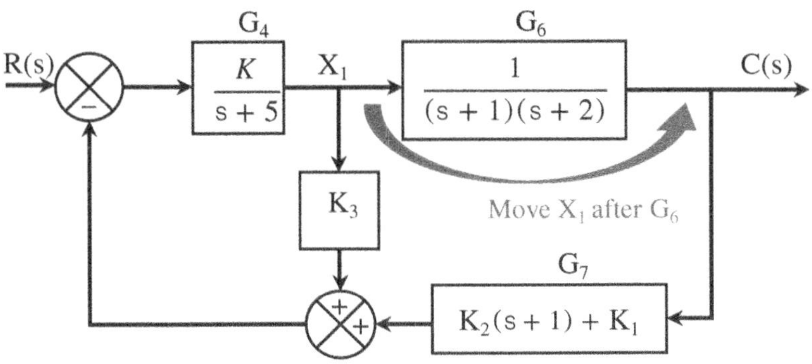

Figure 3.17-b The block diagram after completing steps 3 and 4

5- Moving the takeoff point 'X_1' after G_6 block as illustrated in Fig. 3.17-b gives Fig. 3.17-c.

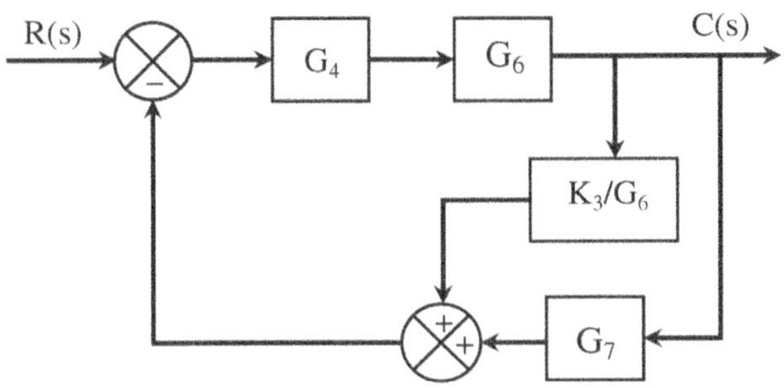

Figure 3.17-c The block diagram after completing step 5

Where $G_4 = KG_1,\ G_6 = G_2G_3,\ G_7 = \dfrac{K_2}{G_3} + K_1$

6- Combine G_4 and G_6 blocks in series.
7- Combine K_3/G_6 and G_7 blocks in parallel results in Fig. 3.17-d.

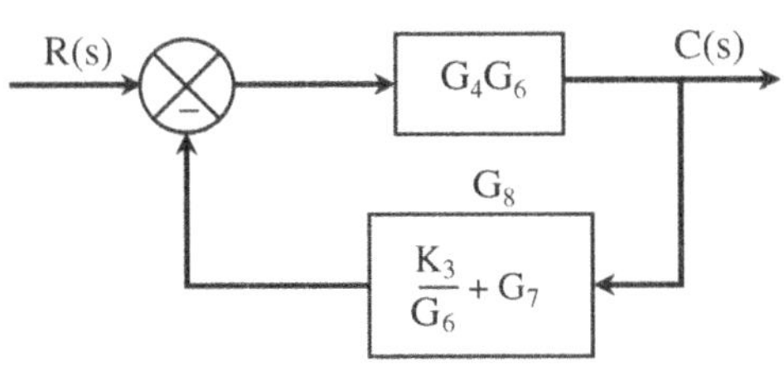

Figure 3.17-d The block diagram after completing steps 6 and 7.

8- Eliminate the feedback loop to get the overall transfer function as shown in Fig. 3.17-e.

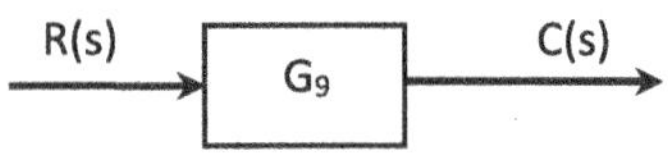

Figure 3.17-e The block diagram after completing step 8

Where $G_9 = \dfrac{G_4G_6}{1 + G_4G_6\left(\frac{K_3}{G_6} + G_7\right)}$

The final transfer function will be:

$$\frac{C(s)}{R(s)} = \frac{KG_1G_2G_3}{1 + KG_1G_2G_3\left(\frac{K_3}{G_2G_3} + \frac{K_2}{G_3} + K_1\right)}$$

$$\frac{C(s)}{R(s)} = \frac{KG_1G_2G_3}{1 + K_3KG_1 + K_2KG_1G_2 + K_1KG_1G_2G_3}$$

The MATLAB code will be:

```
clear
clc
% file name: Ex 3_17.m
%% we suppose values for k, k1, k2 and k3 in order
%  to get results in MATLAB
k=1
k1=10,
k2=2,
k3= 10,
s=tf('s')
G1=1/(s+5)
G2=1/(s+2)
G3=1/(s+1)
G4=series(k,G1)
% moving the takeoff point X2 after G3 block
G5=k2/G3
G6=series(G2,G3)
G7=parallel(G5,k1)
G46=series(G4,G6)
G8=parallel(k3/G6,G7)
disp('the overall transfer function')
G9=feedback(G46,G8)
```

The Second Method:

In this method, take the output after each block and write it in an equation form starting from the output C and substituting the variables until reaching the input R. Placing letters after each block instead of writing the whole equation makes your block diagram with fewer annotations as shown in Fig. 3.13-f.

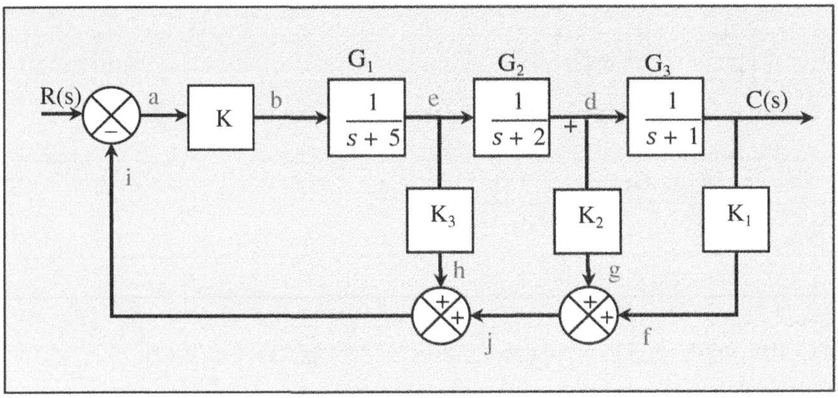

Figure 3.17-f Writing letters at the end of each block instead of equations.

Put letters a, b, e, d, f, g, j, h, i after each block.
We will start from C the output:

$$C = dG_3 \tag{3.11}$$

From Eq. (3.11), we can get d in terms of C, $d = \frac{C}{G_3}$

$$d = eG_2 \tag{3.12}$$

From Eq. (3.12), we can get e in terms of C

$$e = \frac{d}{G_2} = \frac{C}{G_2G_3}$$

$$e = bG_1 \tag{3.13}$$

$$b = aK \tag{3.14}$$

$$a = R - i \tag{3.15}$$

$$h = eK_3 \tag{3.16}$$

$$g = dK_2 \tag{3.17}$$

$$f = CK_1 \tag{3.18}$$

$$j = g + f = dK_2 + CK_1 = \frac{C}{G_3}K_2 + CK_1 = C\left(\frac{K_2}{G_3} + K_1\right)$$

$$i = h + j = eK_3 + C\left(\frac{K_2}{G_3} + K_1\right) = \frac{C}{G_3G_2}K_3 + C\left(\frac{K_2}{G_3} + K_1\right)$$

$$C = dG_3$$

$$= eG_2\ G_3$$

$$= bG_1\ G_2G_3$$

$$= aK\,G_1\ G_2G_3$$
$$= (R - i)K\,G_1\ G_2G_3$$
$$C = \left(R - \left[\frac{CK_3}{G_3G_2} + C\left(\frac{K_2}{G_3} + K_1\right)\right]\right) KG_1G_2G_3$$
$$C = \left(R - C\left[\frac{K_3}{G_3G_2} + \frac{K_2}{G_3} + K_1\right]\right) KG_1G_2G_3$$
$$C = RKG_1G_2G_3 - C\left[\frac{K_3}{G_3G_2} + \frac{K_2}{G_3} + K_1\right] KG_1G_2G_3$$

By transferring the C term from the right side to the left side to get the final C(s)/R(s):

$$\frac{C(s)}{R(s)} = \frac{KG_1G_2G_3}{1 + K_3KG_1 + K_2KG_1G_2 + K_1KG_1G_2G_3}$$

Example 3.18 Find C(s)/R(s) for the block diagram of Fig. 3.18 [7, p: 215, Fig. 9.12].

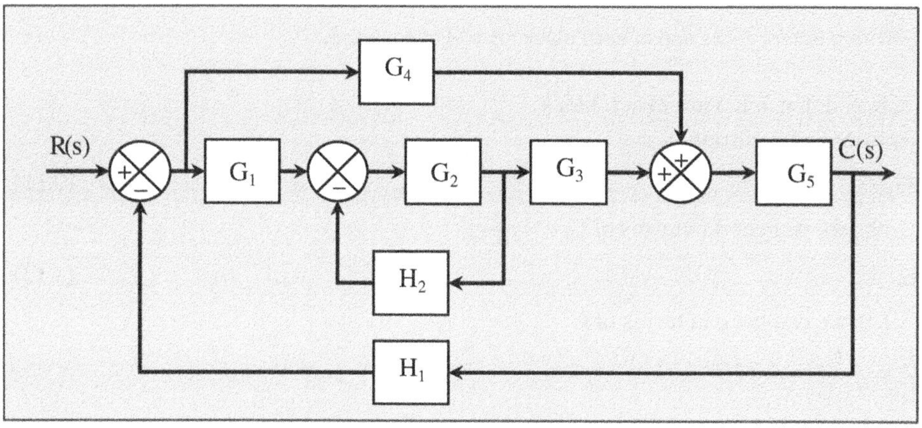

Figure 3.18 The original block diagram of Example 3.18.

Solution:

1- Eliminating the negative feedback loop formed by block H_2 gives G_6 as shown in Fig. 3.18-a.

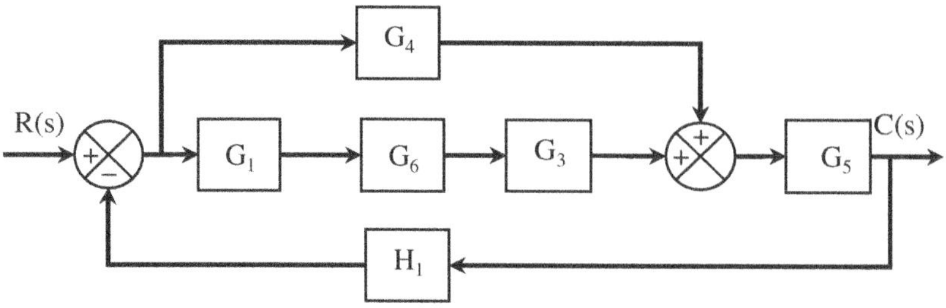

Figure 3.18-a The block diagram after completing step 1

Where $G_6 = \dfrac{G_2}{1 + G_2H_2}$

2- Combining G_1, G_6, and G_3 blocks in series, then combining the result in parallel with G_4 gives G_7 as shown in Fig. 3.18-b.

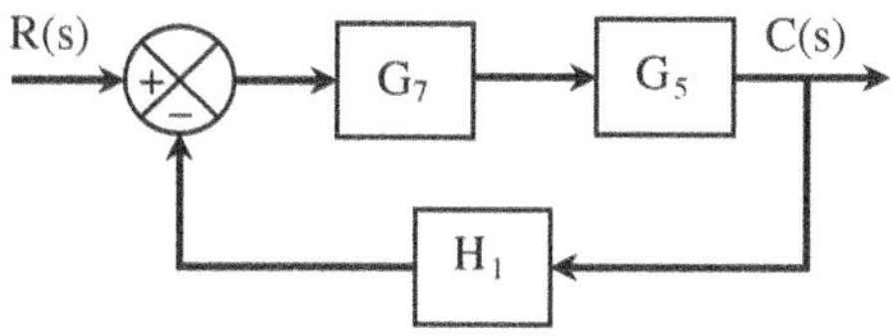

Figure 3.18-b The block diagram after completing step 2

Where $G_7 = G_1G_6G_3 + G_4$

3- Combining G_7 and G_5 blocks in series and then eliminating the negative feedback loop formed by block H_1 gives G_8 as shown in Fig. 3.18-c.

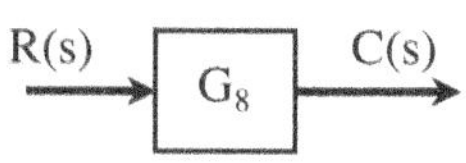

Figure 3.18-c The block diagram after completing step 3

Where $G_8 = \dfrac{G_7G_5}{1 + G_7G_8H_1}$

The MATLAB code will be:

```
clear
clc
% file name: Ex 3_18.m
% define all numerators and denominators for all G's and H's blocks
nG1=[    ]; dG1=[    ];
nG2=[    ]; dG2=[    ];
nG3=[    ]; dG3=[    ];
nG4=[    ]; dG4=[    ];
nG5=[    ]; dG5=[    ];
nH1=[    ]; dH1=[    ];
nH2=[    ]; dH2=[    ];
% feedback G2 and H2
[nG6,dG6]=feedback(nG2,dG2,nH2,dH2,-1);
% G1, G6 and G3 in series
[nG16,dG16]=series(nG1,dG1,nG6,dG6);
[nG163,dG163]=series(nG16,dG16,nG3,dG3);
% parallel G163 and G4
[nG7,dG7]=parallel(nG163,dG163,nG4,dG4);
% G7 and G5 in series
[nG75,dG75]=series(nG7,dG7,nG5,dG5);
% feedback G75 and H1
[nG8,dG8]=feedback(nG75,dG75,nH1,dH1,-1);
disp('the overall system transfer function')
G8=tf(nG8,dG8)
```

Activity:

- Write a MATLAB code for these reduction steps in another way.
- Try to solve this block diagram by using another method.

Example 3.19 Find the overall transfer function for the block diagram of Fig. 3.19 [7, p: 326, Fig. P9.9].

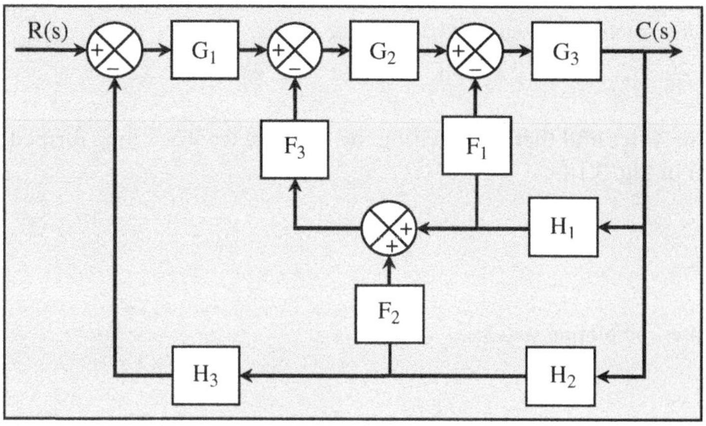

Figure 3.19 The original block diagram of Example 3.19.

Solution:

Two methods are presented here to solve the block diagram of Fig. 3.19; the first one shows a pseudocode for the reduction steps, while the second one shows the equation form.

The First Method:

1- The block diagram of Fig. 3.19 can be redrawn as shown in Fig. 3.19-a.
2- Combine blocks F_1 and H_1 in series, then eliminate the feedback loop formed by block F_1H_1 as illustrated with the long dash line in Fig. 3.19-a.
3- Combine blocks F_2 and H_2 in series, then combine block H_1 in parallel with block F_2H_2 as illustrated with the round dot line in Fig. 3.19-a.
4- Combine the result of step 3 in series with block F_3.
5- Combine blocks H_3 and H_2 in series as illustrated with the bold line in Fig. 3.19-a.

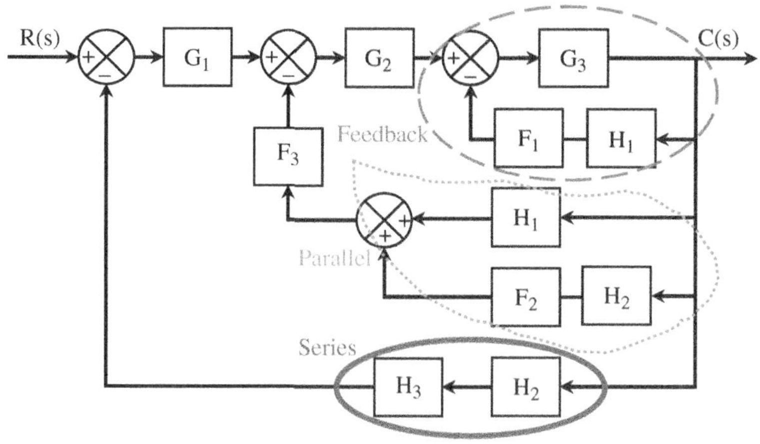

Figure 3.19-a The block diagram that illustrates steps 1 to 5.

6- After the completion of these steps, the block diagram of Fig. 3.19-a will be as shown in Fig. 3.19-b.

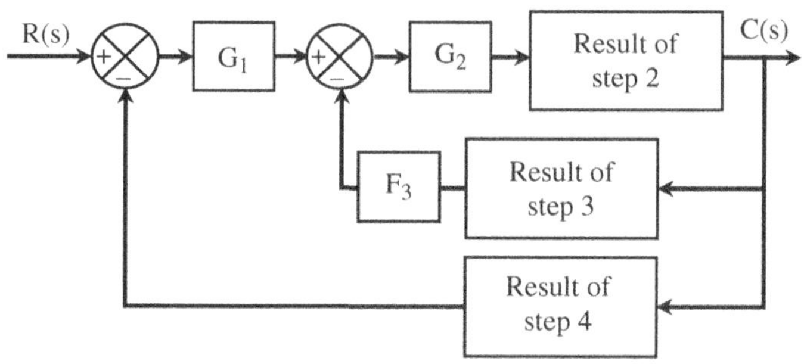

Figure 3.19-b The block diagram after completing step 6.

7- After continuing the steps of reduction, the pseudocode for MATLAB will be:

1- Clear.
2- clc.
3- Name your file.
4- Series (F_1 and H_1).
5- Eliminate the feedback loop formed by F_1H_1.
6- Series (F_2 and H_2).
7- Parallel (H_1 and F_2H_2).
8- Combine the result of line 7 in series with F_3 block.
9- Series (H_3 and H_2).
10- Series (G_2 and result of step 2) as shown in Fig. 3.19-b.
11- Eliminate the negative feedback formed by the result of line 8.
12- Series (G_1 and results of line 11).
13- Eliminate the negative feedback formed by the result of step 4 as shown in Fig. 3.19-b.

Activity:

Write the MATLAB code for these block diagram reduction steps.

The Second Method:

By writing the output after each block, Fig. 3.19 can be redrawn as Fig. 3.19-c.

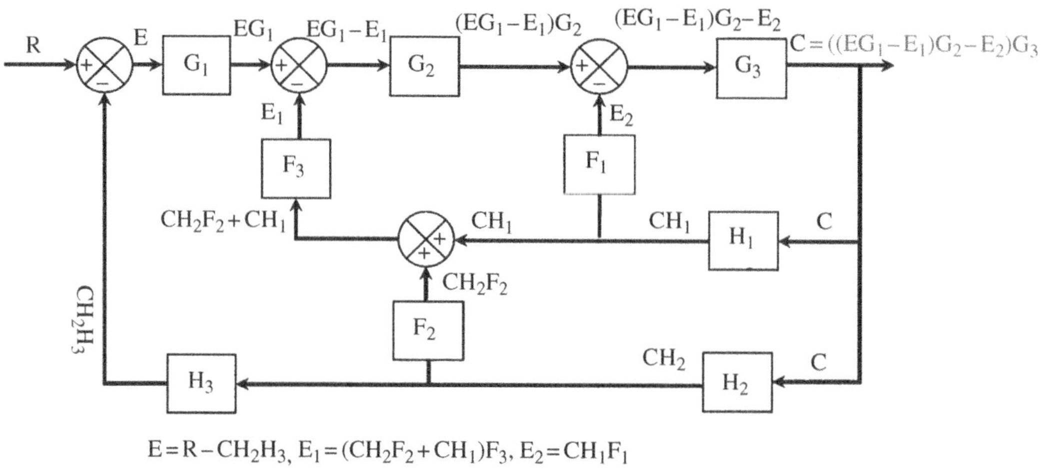

Figure 3.19-c The outputs after each block

Start from the output C

$$C = ((EG_1 - E_1)G_2 - E_2)G_3 \tag{3.19}$$
$$= ((EG_1G_2 - E_1G_2) - E_2)G_3$$
$$= EG_1G_2G_3 - E_1G_2G_3 - E_2G_3$$
$$E = R - CH_2H_3 \tag{3.20}$$
$$E_1 = (CH_2F_2 + CH_1)F_3 \tag{3.21}$$
$$E_2 = CH_1F_1 \tag{3.22}$$

Substitute for E, E_1, and E_2 in Eq. (3.19) to get:

$$C = (R - CH_2H_3)G_1G_2G_3 - (CH_2F_2 + CH_1)F_3G_2G_3 - CH_1F_1G_3$$

After simplifications, the overall transfer function will be:

$$\frac{C(s)}{R(s)} = \frac{G_1G_2G_3}{1 + (G_1H_2H_3 + H_2F_2F_3 + H_1F_3)G_2G_3 + G_3H_1F_1}$$

Example 3.20 Prove that the transfer function Y(s)/U(s) for the block diagram of Fig. 3.20 is equal to:

$$\frac{Y(s)}{U(s)} = \frac{1/2}{s+2} + \frac{-1}{s+3} + \frac{1/2}{s+4}$$

[7, p: 158, Fig. 6-12].

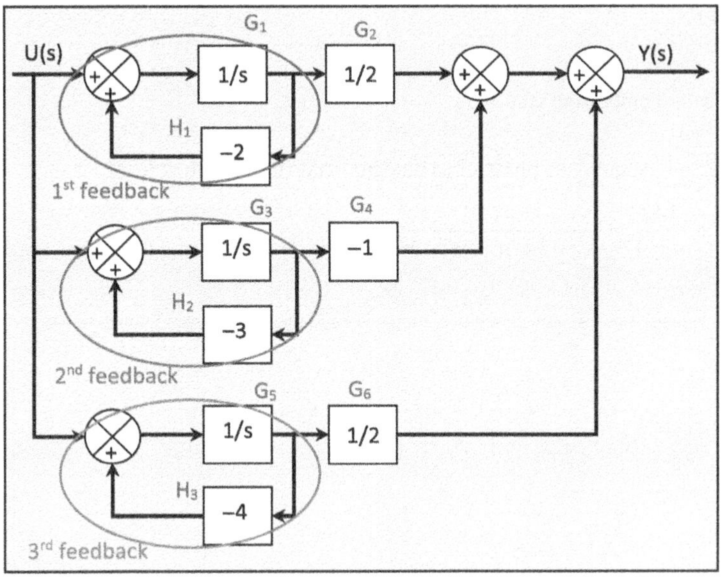

Figure 3.20 The original block diagram of Example 3.20.

Solution:

1- Eliminate the three feedback loops and combine the two summing points as illustrated in Fig. 3.20, then your figure should look like Fig. 3.20-a.

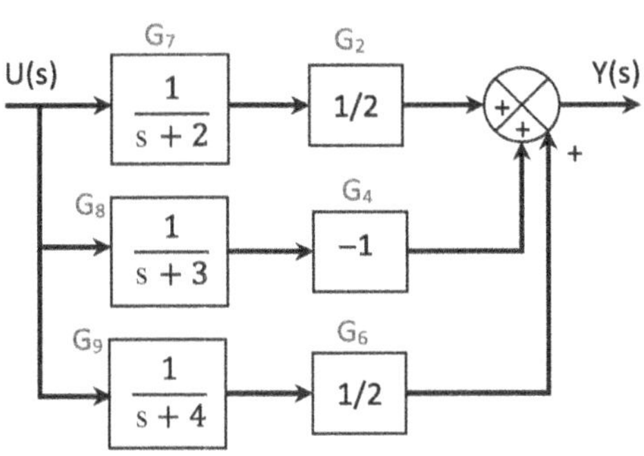

Figure 3.20-a The block diagram after completing step 1.

2- Combine blocks $\frac{1}{s+2}$ and $\frac{1}{2}$ in series, combine blocks $\frac{1}{s+3}$ and -1 in series, and combine blocks $\frac{1}{s+4}$ and $\frac{1}{2}$ in series. The result will be as shown in Fig. 3.20-b.

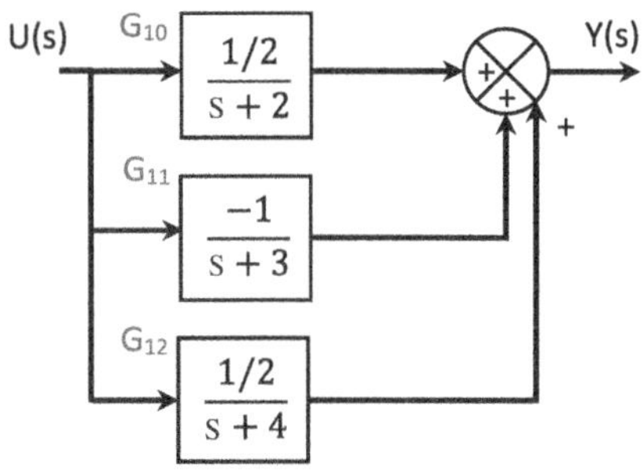

Figure 3.20-b The block diagram after completing step 2.

3- Combine the three blocks $\frac{1/2}{s+2}$, $\frac{-1}{s+3}$ and $\frac{1/2}{s+4}$ in parallel to get the final transfer function.

$$\frac{Y(s)}{U(s)} = \frac{1/2}{s+2} + \frac{-1}{s+3} + \frac{1/2}{s+4} = \frac{1}{s^3 + 9s^2 + 26s + 24}$$

The MATLAB code will be:

```
clear
clc
% file name: Ex 3_20.m
s=tf('s');
G1=1/s; G2=1/2;
H1=-1*2;
G3=1/s; G4=-1;
H2=-1*3;
G5=1/s; G6=1/2;
H3= -1*4;
% elimination of three feedback loops
G7=feedback(G1,H1,+1)
G8=feedback(G3,H2,+1)
G9=feedback(G5,H3,+1)
% three series groups
G10=series(G7,G2)
G11=series(G8,G4)
G12=series(G9,G6)
% three parallel group
G13=parallel(G10,G11)
disp('the overall transfer function Y(s)/U(s)=')
parallel(G13,G12)
```

After Execution:

```
G7 =
      1
    -----
    s + 2
```

```
G8 =

    1
  -----
  s + 3

G9 =

    1
  -----
  s + 4

G10 =

   0.5
  -----
  s + 2

G11 =

   -1
  -----
  s + 3

G12 =

   0.5
  ------
  s + 4

G13 =

  -0.5 s - 0.5
  -------------
  s^2 + 5 s + 6

the overall transfer function Y(s)/U(s)=

            1
  -------------------------
  s^3 + 9 s^2 + 26 s + 24
```

Activity:

Write the MATLAB code in another way.

Example 3.21 Find the closed-loop transfer function C(s)/R(s) for the system of Fig. 3.21.

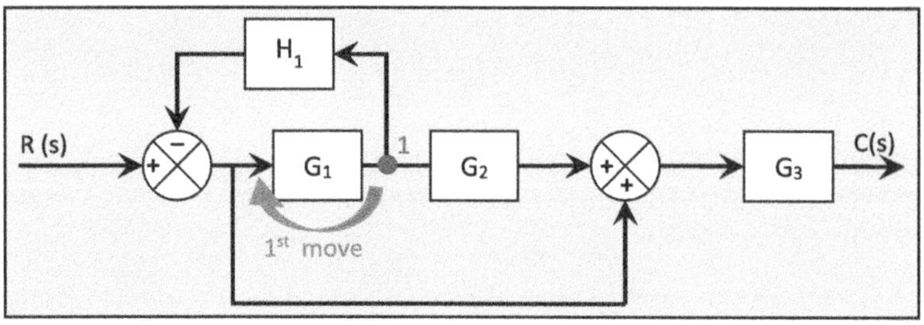

Figure 3.21 The original block diagram of Example 3.21.

Solution:

Two methods are presented to solve this example; the first one is based on using the block diagram reduction rules, while the second is based on writing the outputs after each block in equation form and then solving them.

The First Method:

1- Moving the takeoff point "1" of block H_1 before block G_1 as demonstrated as the first move in Fig. 3.21 gives Fig. 3.21-a.

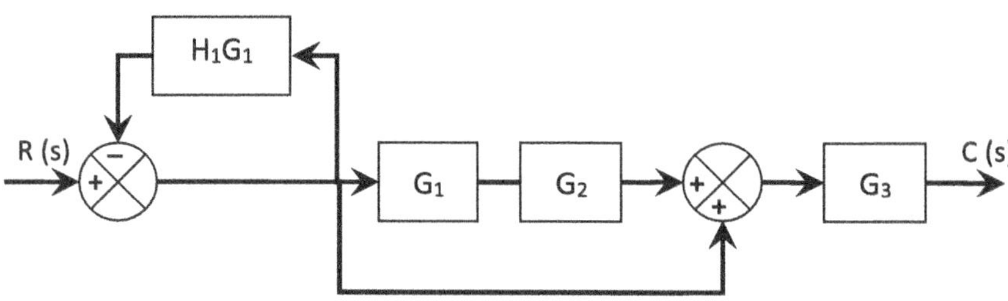

Figure 3.21-a The block diagram after completing step 1.

The MATLAB code for this case will be:

```
clear
clc
% file name: Ex3_21.m
s=tf('s');
% suppose you have
G1=1/(s+1); G2=2/(s+2);G3=4*s; H1=-2;
% Move a takeoff point before G1 block
G4=H1*G1
```

After Execution:

```
G4 =

   -2
  -----
  s + 1
```

2- Eliminate the negative feedback loop formed by block H_1G_1. Combining blocks G_1 and G_2 in series gives Fig. 3.21-b.

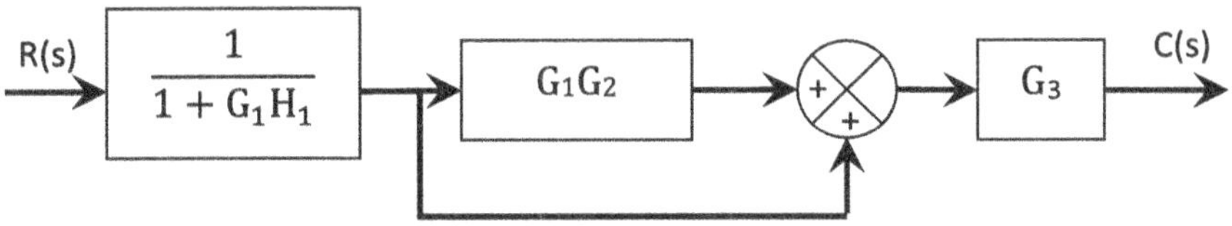

Figure 3.21-b The block diagram after completing step 2.

```
% eliminate the -ve feedback formed by H1G1
G5=feedback(1,G4,+1)
% G1 and G2 in series
G6=series(G1,G2)
```

After Execution:

```
G5 =

  s + 1
  -----
  s + 3

G6 =

        2
  -------------
  s^2 + 3 s + 2
```

3- Combining block G_1G_2 with the unity path in parallel results in Fig. 3.21-c.

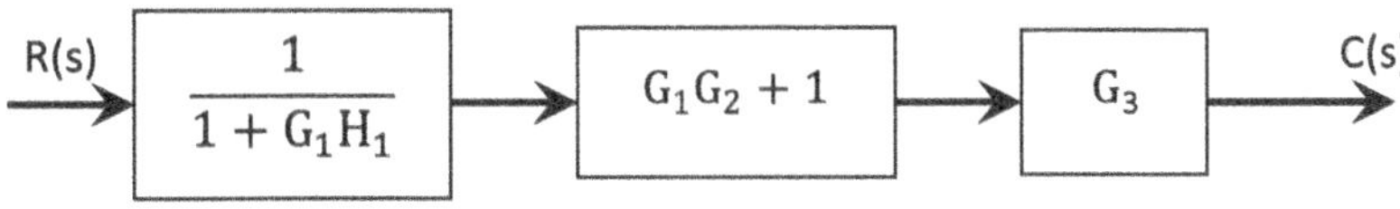

Figure 3.21-c The block diagram after completing step 3.

```
% combining G6 in parallel with 1
  G7=parallel(G6,1)
```

After Execution:

```
G7 =

       s^2 + 3 s + 4
      ---------------
       s^2 + 3 s + 2
```

4- Combining the three blocks of Fig. 3.21-c in series gives the overall system transfer function as shown in Fig. 3.21-d.

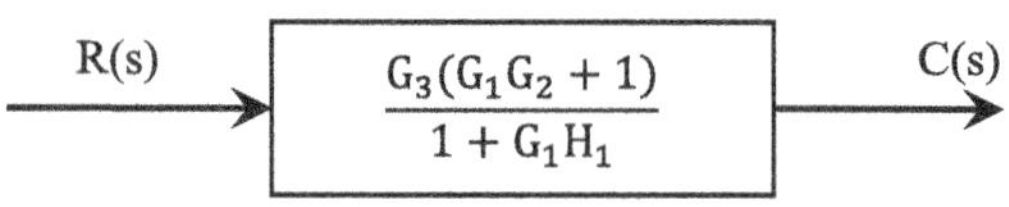

Figure 3.21-d The block diagram after completing step 4.

$$\frac{C(s)}{R(s)} = \frac{G_3(G_1G_2 + 1)}{1 + G_1H_1}$$

```
disp('the overall transfer function Y(s)/U(s)=')
G8=series(G5,G7);
G9=series(G8,G3)
```

After Execution:

```
G9 =

  4 s^4 + 16 s^3 + 28 s^2 + 16 s
  --------------------------------
      s^3 + 6 s^2 + 11 s + 6
```

The Second Method:

The output after each block is shown in Fig. 3.21-e.

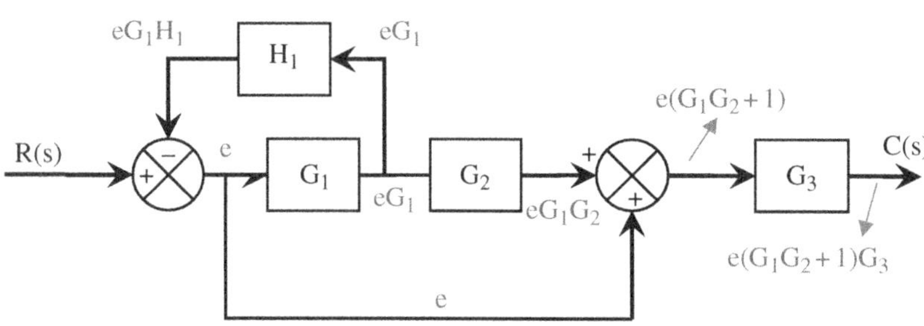

Figure 3.21-e The block diagram of Example 3.21 with the output after each block.

Two equations are acquired: one for the error (e) and one for the output (c).

$$e = R - eG_1H_1 \tag{3.23}$$

Then

$$e = \frac{R}{1 + G_1H_1} \tag{3.23'}$$

$$c = e(G_1G_2 + 1)G_3 \tag{3.24}$$

Substitute Eq. (3.23′) into Eq. (3.24) to get,

$$c = \frac{R}{1 + G_1H_1}(G_1G_2 + 1)G_3$$

Then C(s)/R(s) will be:

$$\frac{C(s)}{R(s)} = \frac{G_3(G_1G_2 + 1)}{1 + G_1H_1}$$

And it is the same answer as in the first method.

For More Practice:

See Ex. 6.8 on page 155 [4].

Example 3.22 Find the closed-loop transfer function C(s)/R(S) for the system of Fig. 3.22 [8, p: 261, Fig. 2-27].

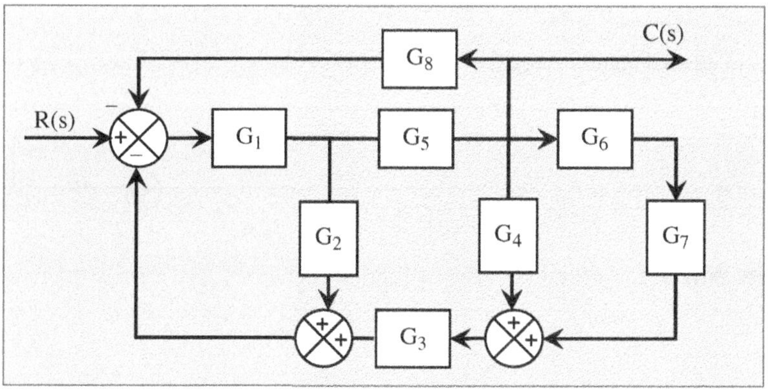

Figure 3.22 The original block diagram of Example 3.22.

Solution:

Two methods are presented to solve this example; the first one is based on using the rules of block diagram reduction with the MATLAB commands after each stage, while the second one is based on writing the outputs after each block in equation form and then solving them.

The First Method:

1- Start your MATLAB code with `Clear` and `clc` then enter all G's transfer functions.

```
clear
  clc
  % % file name: Ex 3_22.m
  s=tf('s');
  % Define all G's and H's blocks
  G1=1/(s+10); G2=1/(s+1);
  G3=1; G4=(s+1)/6; G5=s;
  G6=2; G7=(s+1)/(s+2); G8=1;
```

2- After combining blocks G_6 and G_7 in series, Fig. 3.22 can be redrawn as shown in Fig. 3.22-a.

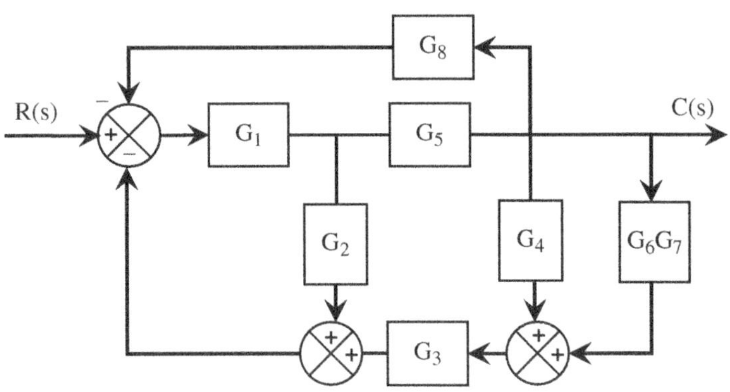

Figure 3.22-a The block diagram after completing step 2.

```
G67=series(G6,G7)
```

After Execution, The Results Will Be:

```
G67 =
    2 s + 2
   ---------
    s + 2
```

3- Combining blocks G_4 and G_6G_7 in parallel gives block G_9 and the block diagram will be as shown in Fig. 3.22-b.

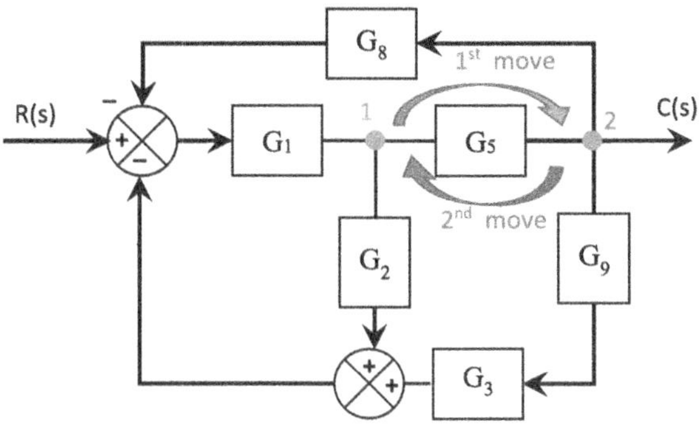

Figure 3.22-b The block diagram after completing step 3.

Where $G_9 = G_4 + G_6G_7$

```
G9=parallel(G4,G67)
```

After Execution, The Results Will Be:

```
G9 =
    s^2 + 15 s + 14
   -----------------
       6 s + 12
```

4- Combining blocks G_3 and G_9 in series gives G_3G_9. Then moving point "1" after block G_5 as illustrated as the 1st move in Fig. 3.22-b gives G_2/G_5. Then moving the takeoff point "2" before block G_5 as demonstrated as the 2nd move in Fig. 3.22-b gives G_8G_5. After the completion of these steps, the block diagram will be as shown in Fig. 3.22-c.

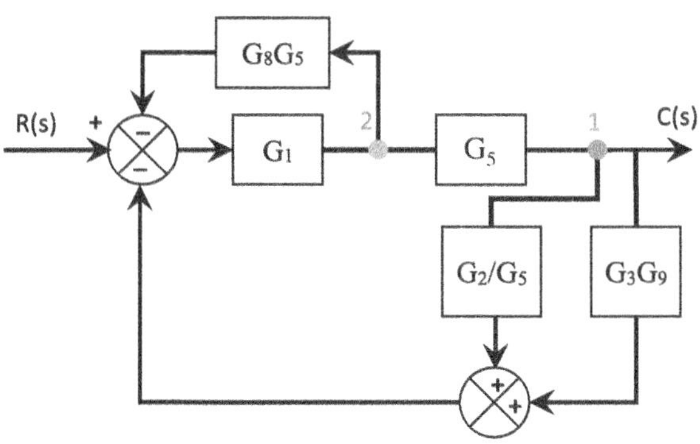

Figure 3.22-c The block diagram after completing step 4.

```
G39=series(G3,G9)   % G39 is G3G9

%% The 1st move and how to get G2/G5
% the first method to get G2/G5
num2=[1]; den2=[1  1];
num5=[1  0]; den5=[1];
n=conv(num2,den5);d=conv(den2,num5);
G2o5=tf(n,d)   % G2o5 is G2/G5
%% another process to find G2/G5
% first, get 1/G5
num5=[1  0]; den5=[1];
d=tf(den5,num5);       %d=1/G5
% second, multiply G2 by 1/G5
G2oG5=series(G2,d) % G2oG5 is G2/G5
% Note that G2oG5 is the same as G2o5 we just
% write it like this in order to recognize the
% two methods
% The 2nd move and how to get G8G5
G85=series(G8,G5)
```

After Execution, The Results Will Be:

```
G39 =

        s^2 + 15 s + 14
        ----------------
           6 s + 12

G2o5 =

           1
       --------
       s^2 + s
G2oG5 =
           1
       --------
       s^2 + s
 G85 =
       s
```

5- Combining blocks G_2/G_5 and G_3G_9 in parallel gives G_{10}. Before eliminating the negative feedback formed by block G_8G_5, Fig. 3.22-c can be redrawn in a more understandable form as shown in Fig. 3.22-d.

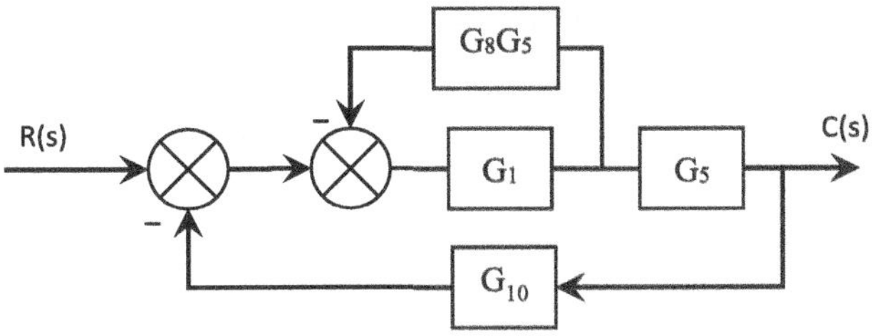

Figure 3.22-d The block diagram after completing step 5

Where $G_{10} = \frac{G_2}{G_5} + G_3G_9$

```
G10=parallel(G2o5,G39)
```

After Execution, The Results Will Be:

```
G10 =
   s^4 + 16 s^3 + 29 s^2 + 20 s + 12
 -----------------------------------
        6 s^3 + 18 s^2 + 12 s
```

6- Eliminating the negative feedback formed by block G_8G_5 gives G_{11} as shown in Fig. 3.22-e.

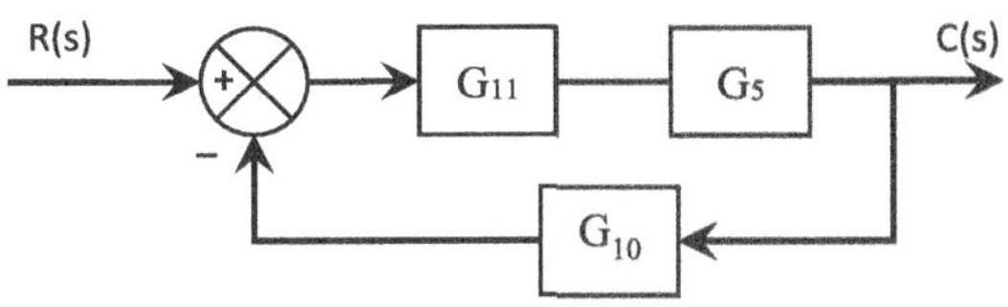

Figure 3.22-e The block diagram after completing step 6

Where $G_{11} = \frac{G_1}{1 + G_1G_8G_5}$

```
% Eliminate the -ve feedback loop formed by G85
G11=feedback(G1,G85,-1)
```

After Execution, The Results Will Be:

```
G11 =

     1
  --------
  2 s + 10
```

7- Combining blocks G_{11} and G_5 in series, then eliminating the negative feedback formed by block G_{10}, results in C(s)/R(s) as shown in Fig. 3.22-f.

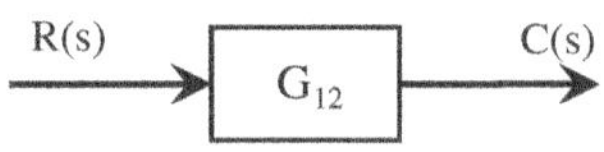

Figure 3.22-f The block diagram after completing step 7

Where $G_{12} = \dfrac{G_{11}G_5}{1 + G_{11}G_5G_{10}}$

After the substitution of G_{11} and G_{10}, the overall transfer function will be:

$$\frac{C(s)}{R(s)} = G_{12} = \frac{G_1G_5}{1 + G_1G_2 + G_1G_5(G_3G_4 + G_3G_6G_7 + G_8)}$$

```
% Overall transfer function C/R=G12=G11G5
     /(1+G11G5G10)
a= series(G11,G5);
G12=feedback(a,G10,-1)
```

After Execution, The Results Will Be:

```
G12 =
           6 s^4 + 18 s^3 + 12 s^2
  ------------------------------------------
  s^5 + 28 s^4 + 125 s^3 + 224 s^2 + 132 s
```

The Second Method:

The output after each block is shown in Fig. 3.22-g.

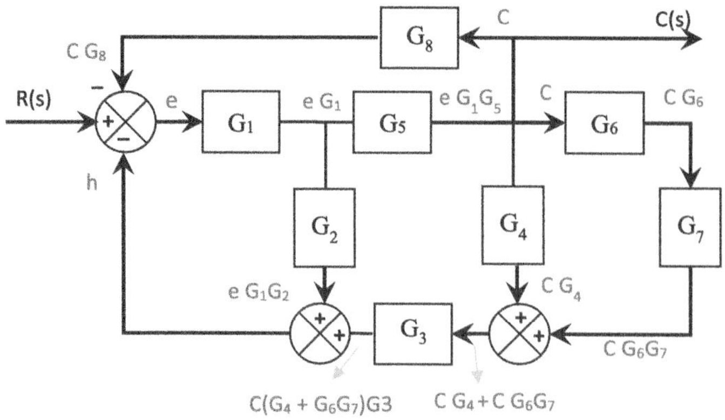

Figure 3.22-g The block diagram of Example 3.22 with the outputs after each block

Where $h = G_3C(G_4 + G_6G_7) + eG_1G_2$

The error equation:

$$e = R - h - CG_8 \tag{3.25}$$

The output equation:

$$C = eG_1G_5 \tag{3.26}$$

Now substitute the equation of h in Eq. (3.25) to get e:

$$e = \frac{R - C(G_3(G_4 + G_6G_7) + G_8)}{1 + G_1G_2} \tag{3.27}$$

Then substitute Eq. (3.27) in (3.26) to get C:

$$C = \frac{R - C(G_3(G_4 + G_6G_7) + G_8)}{1 + G_1G_2} G_1G_5 \tag{3.28}$$

After simplifications C(s)/R(s) will be:

$$\frac{C(s)}{R(s)} = \frac{G_1G_5}{1 + G_1G_2 + G_1G_5(G_3G_4 + G_3G_6G_7 + G_8)}$$

Example 3.23 Find the closed-loop transfer function C(s)/R(s) for the system of Fig. 3.23.

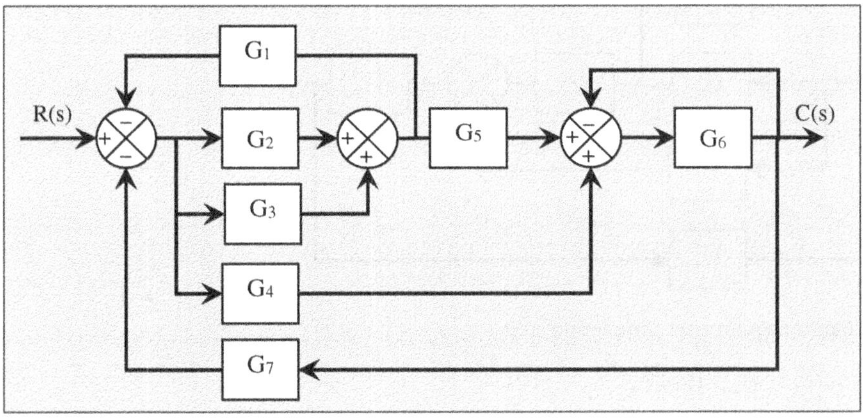

Figure 3.23 The original block diagram of Example 3.23.

Solution:

The reduction stages are explained here with their pseudocode:

1- In cases like one line is distributed to blocks G_3 and G_4 and multi-input summing points (the first and third), Fig. 3.23 can be redrawn as shown in Fig. 3.23-a to be more understandable.

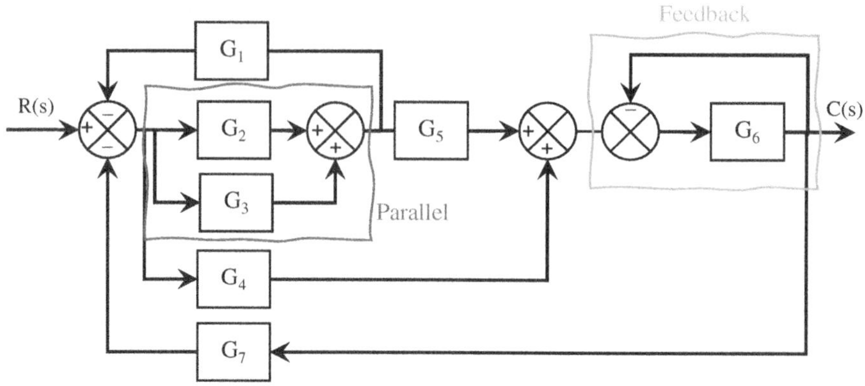

Figure 3.23-a The block diagram of Example 3.23 in more understandable form.

2- Combine blocks G_2 and G_3 in parallel and eliminate the negative unity feedback as demonstrated in Fig. 3.23-a. Your block diagram will be as shown in Fig. 3.23-b.

the pseudocode will be:

1- Combine blocks G_2 and G_3 in parallel then eliminate the negative unity feedback loop.

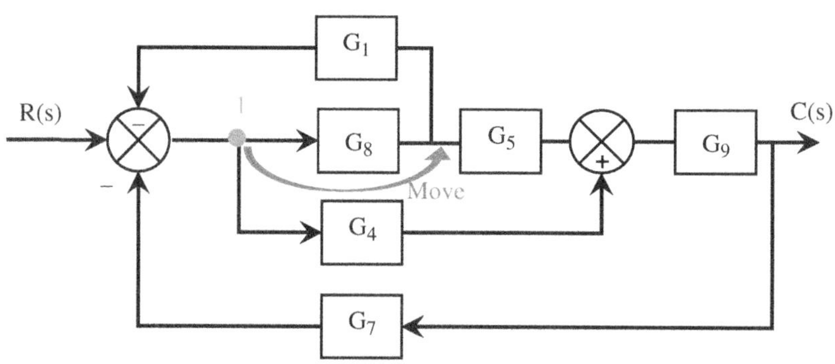

Figure 3.23-b The block diagram after completing step 2

Where $G_9 = \frac{G_6}{1+G_6}$, $G_8 = G_2 + G_3$

3- Moving the takeoff point "1" after block G_8 as illustrated in Fig. 3.23-b gives Fig. 3.23-c.

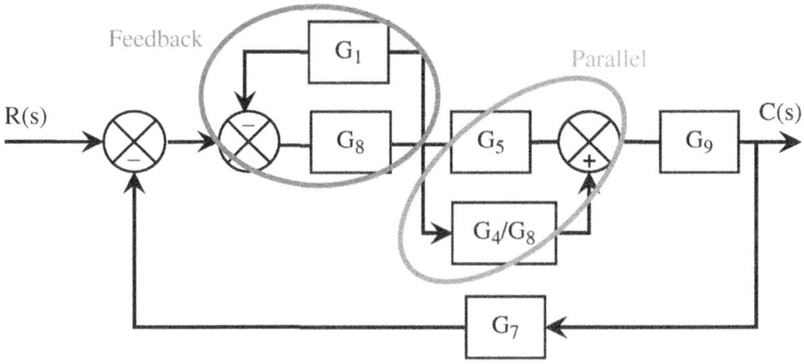

Figure 3.23-c The block diagram after completing step 3.

2- Moving the takeoff point "1" after block G_8 results in G_4/G_8

4- Eliminating the negative feedback formed by block G_1 and combining blocks G_5 and G_4/G_8 in parallel as illustrated in Fig. 3.23-c, results in Fig. 3.23-d.

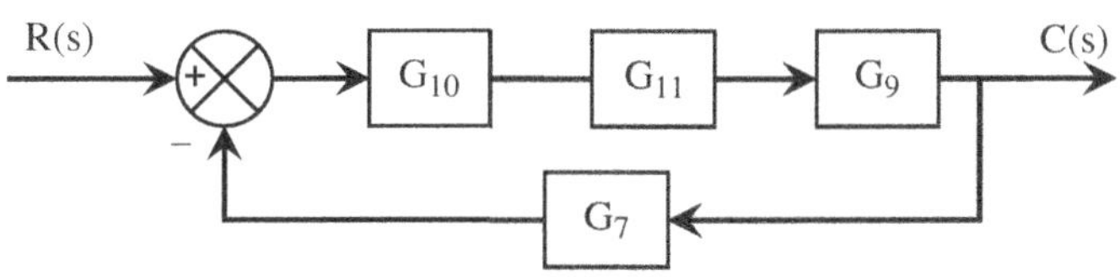

Figure 3.23-d The block diagram after completing step 4

Where $G_{10} = \dfrac{G_8}{1+G_8G_1}, G_{11} = G_5 + \dfrac{G_4}{G_8}$

3- Eliminate the negative feedback loop formed by G_1. 4- Blocks G_5 and G_4/G_8 in parallel.

Activity:

- Reduce the block diagram of Fig. 3.23-d to obtain the final transfer function.
- Write the rest pseudocode statements.

For More Practice:

See Ex. 3-1-5 on page 114, Chapter 3 [9].

Example 3.24 Obtain the overall transfer function for the system of Fig. 3.24. The idea of this block diagram was taken from Fig. 6.19 on page 174 with different symbols [1].

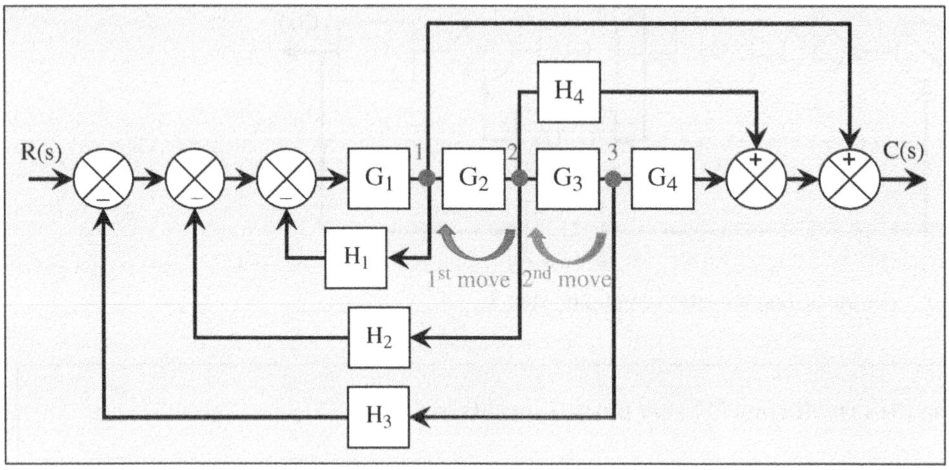

Figure 3.24 The original block diagram of Example 3.24

$$G_1 = \frac{1}{s},\ G_2 = \frac{1}{s},\ G_3 = \frac{1}{s},\ G_4 = 2,\ H_1 = 9,\ H_2 = 26,\ H_3 = 24,\ H_4 = 3$$

Solution:

Two ways are presented to solve this example; the first one focuses on combining parallel paths and showing MATLAB code and the execution part. The second one focuses on eliminating the feedback loop. This book has hints for the second way and leaves the rest to students.

The First Way: Combining Parallel Loops:

1- Don't eliminate the feedback loop formed by block H_1 at the beginning; wait until doing the following steps.
2- Moving the takeoff point "2" of block H_2 before block G_2 as demonstrated as the 1st move in Fig. 3.24 gives Fig. 3.24-a.
3- Moving the takeoff point "3" of block H_3 before block G_3 as demonstrated as the 2nd move in Fig. 3.24 gives Fig. 3.24-a.

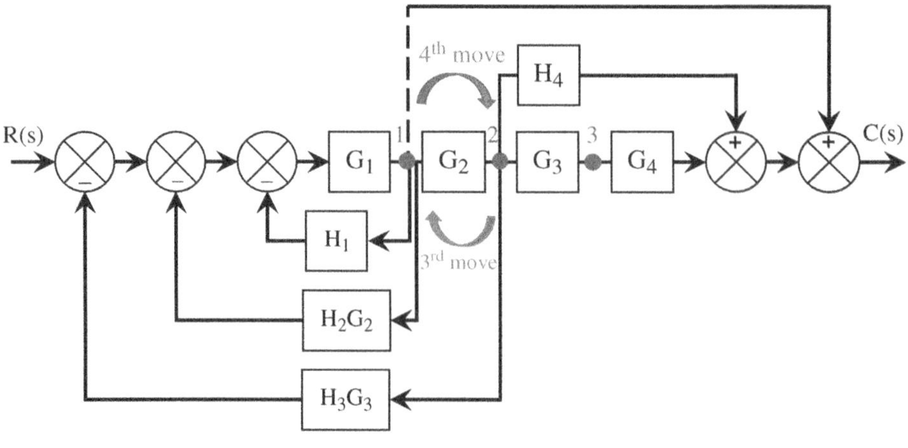

Figure 3.24-a The block diagram after completing steps 2 and 3

4- Moving the takeoff point "2" of block H_3G_3 before block G_2 as demonstrated as the 3rd move in Fig. 3.24-a gives Fig. 3.24-b.

5- Moving the takeoff point "1" of 1 after block G_2 as demonstrated as the 4th move in Fig. 3.24-a gives $1/G_2$ as shown in Fig. 3.24-b.

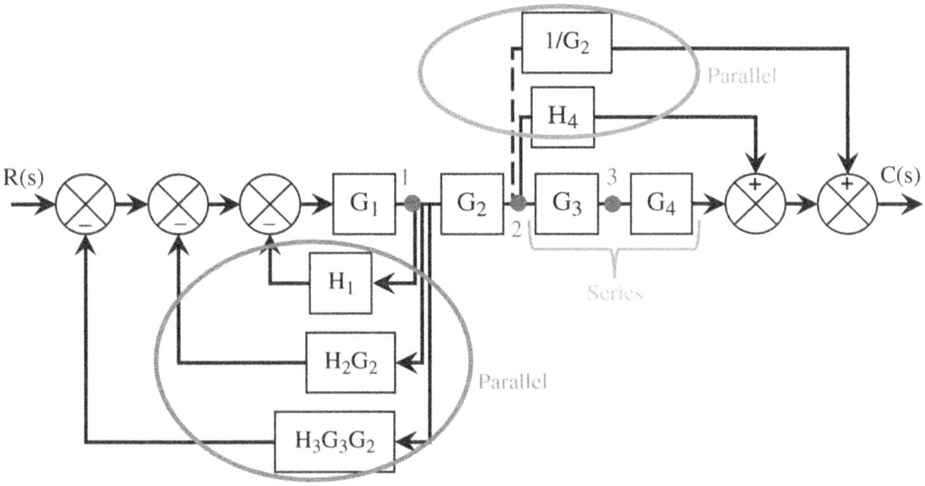

Figure 3.24-b The block diagram after completing step 4 and demonstration of step 5.

6- Combining H_1, H_2G_2, $H_3G_3G_2$ blocks in parallel, $\frac{1}{G_2}$, H_4 blocks in parallel, and G_2 and G_3 blocks in series as illustrated in Fig. 3.24-b gives Fig. 3.24-c.

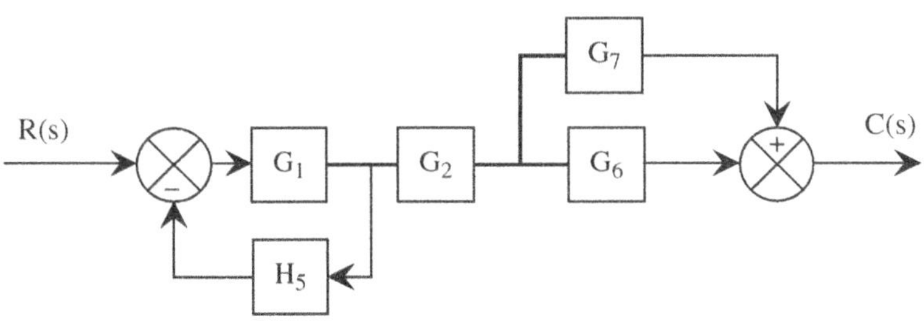

Figure 3.24-c The block diagram after completing step 6

Where $H_5 = H_1 + H_2G_2 + H_3G_3G_2$,

$$G_6 = G_3G_4, \quad G_7 = \frac{1}{G_2} + H_4$$

7- Eliminating the negative feedback loop formed by block H_5 gives G_8. Combining blocks G_6 and G_7 in parallel gives G_9 as shown in Fig. 3.24-d.

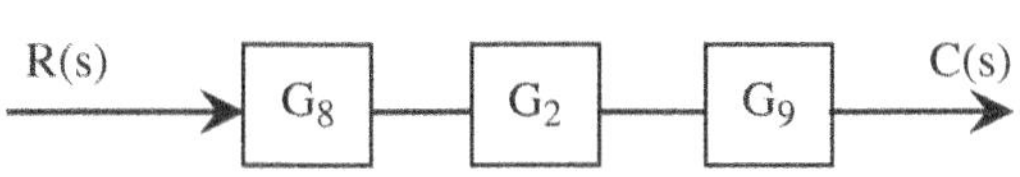

Figure 3.24-d The block diagram after completing step 7

Where $G_8 = \frac{G_1}{1 + G_1H_5}$, $G_9 = G_6 + G_7$

8- To get the overall system transfer function, combine the three blocks (G_8, G_2, and G_9) in series as shown in Fig. 3.24-e.

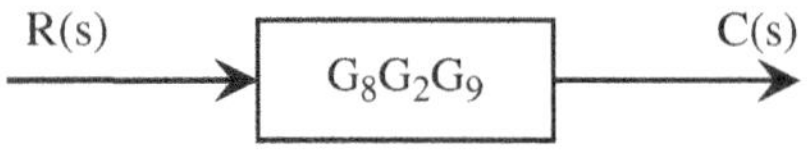

Figure 3.24-e The block diagram after completing step 8.

Then

$$\frac{C(s)}{R(s)} = \frac{G_1G_2(G_6 + G_7)}{1 + G_1H_5}$$

The MATLAB program will be:

```
Clear
Clc
% % file name: Ex 3_24.m
s=tf('s');
% Define all G's and H's blocks
G1=1/s; G2=1/s;
G3=1/s; G4=2;
H1=9; H2=26; H3=24; H4=3;
% Combining parallel loop and Making the required moves
a=series(H2,G2); % the 1st move a=H2G2
b=series(H3,G3); % the 2nd move b=H3G3
c=series(b,G2); % the 3rd move c=H3G3G2
% The 4th move and how to get 1/G2
num=[1 0]; den=1;
d=tf(num,den); %d=1/G2
e=parallel(H1,a);
H5=parallel(e,b)
G7=parallel(d,H4)
G6=series(G3,G4) % series blocks
% Eliminate the -ve feedback loop formed by H5
G8=feedback(G1,H5,-1)
G2
G9=parallel(G6,G7)
% Overall transfer function C/R=G8G2G9
f= series(G8,G2);
G10=series(f,G9)
```

After Execution, The Results Will Be:

```
H5 =

  9 s^2 + 50 s
  ------------
      s^2
```

```
G7 =

  s + 3
G6 =


  2
  -
  s

G8 =         s^2
  ------------------
  s^3 + 9 s^2 + 50 s

G2 =

  1
  -
  s

G9 =

  s^2 + 3 s + 2
  -------------
        s

G10 =

  s^4 + 3 s^3 + 2 s^2
  ---------------------
  s^5 + 9 s^4 + 50 s^3
```

Note:

To get $1/G_2$, write G_2 in numerator and denominator form, and then to get the reciprocal of G_2, you have to place the denominator in the numerator position and place the numerator in the denominator position. In this case, you don't need to use the convolution command in MATLAB.

$$G_2 = \frac{1}{s} \quad \text{the numerator} = 1, \text{the denominator} = s$$

$$\frac{1}{G_2} = \frac{1}{\frac{1}{s}} = \frac{s}{1} \quad \text{the numerator} = s, \text{the denominator} = 1$$

If you have G_3/G_2, you have to use the convolution command in MATLAB when you want G_2/G_3 as it was discussed in Chapter 2.

The Second Way:

1- Start by eliminating the feedback loop formed by H_1.
2- Move the forward path formed by 1 after block G_2.
3- Draw the block diagram after doing 1 and 2, and then complete the rest.

For More Practice:

See Chapter 6 block diagram on pages 141–176 [4].
See Chapter 3 block diagram and signal-flow graphs on pages 104–146 [9].

Review Questions

1 If G_1, G_2, and G_3 blocks are connected in series, draw the block diagram. Write the outputs after each block in equations form and find the overall system transfer function. See Ghosh [p: 144, Ex. 6.1].

2 If G_1, G_2, and G_3 blocks are connected in parallel, draw the block diagram. Write the outputs after each block in equations form and find the overall system transfer function. See Ghosh [p: 145, Ex. 6.2].

3 Write the outputs after each block for the block diagram of Fig. 3.25 then obtain C(s)/R(s) [Ghosh, p: 162, Ex. 6.11].

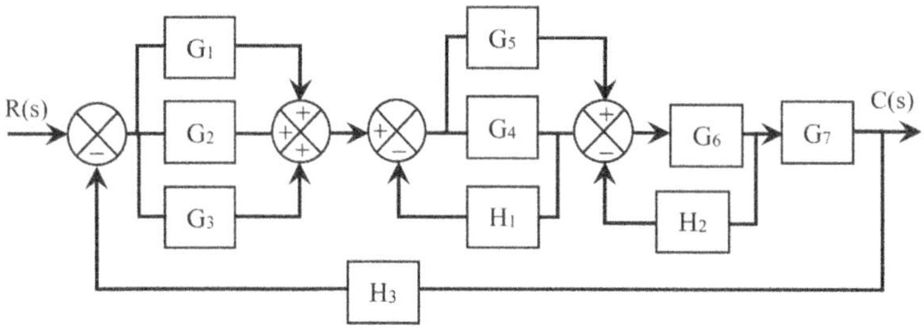

Figure 3.25 The block diagram of Q.3.

4 Simplified the block diagram shown in Fig. 3.26 [Ghosh, p: 172, Fig. 4]

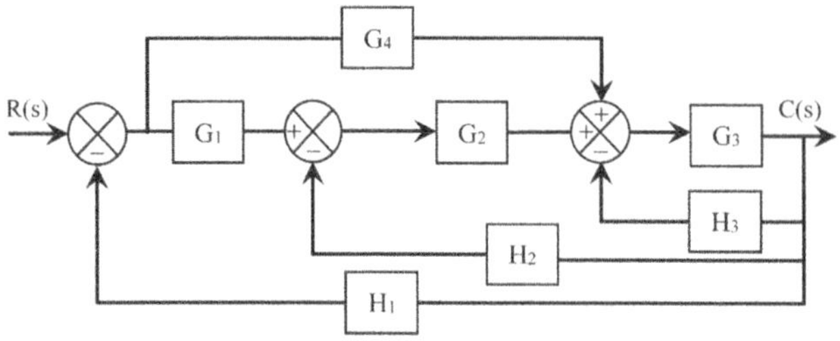

Figure 3.26 The block diagram of Q.4.

5 Find C(s)/R(s) of the block diagram shown in Fig.3.27 by using the block diagram reduction rules.
Write the outputs after each block, then find C(s)/R(s) for the second time and compare the results.

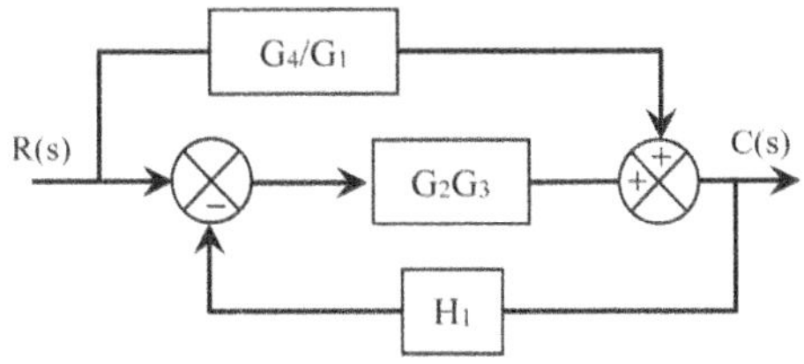

Figure 3.27 The block diagram of Q.5.

4

Creating a Block Diagram in Microsoft® Word

4.1 Introduction

Microsoft® Word is a Word Processor established by Microsoft. It is used to write documents such as letters, reports, resumes, theses, and books. Word enables the user to write text or symbols, draw a shape, insert a photo, draw a flowchart, draw a block diagram, and it has different features that help the user in formatting his/her documents. Word features include the paper size and type, font style, size, color, page header and footer, page numbering, page layout, and more. Word allows users to print their finished files or save them in PDF form.

This chapter focuses on creating a block diagram in Microsoft Word. In the beginning, the reader will learn how to find the components of the block diagram. Next, two methods of creation are described. The first one is based on opening a new canvas and working inside this canvas, while the second is based on working on a blank document. Alignment ways are then discussed. Finally, examples are available to illustrate and demonstrate the discussed methods.

Learning Objectives:

- To learn how to draw a block diagram in Microsoft Word.
- To learn different methods for drawing a block diagram in Microsoft Word.

4.2 Block Diagram Construction

Any block diagram consists of the following parts or components:

- Summing point.
- Block.
- Connectors or arrows.

They can be represented in Microsoft Word as shown in Table 4.1.

Block Diagram Implementation for Control Systems: Including MATLAB® Programming, Microsoft® Word and PowerPoint Presentation, First Edition. Ruba Al-Mulla Hummadi.

Table 4.1 Block diagram components.

Part of a block diagram	Microsoft Word representation	Where to find the shape in Microsoft Word
Summing point		Flowchart summing junction from Shapes menu.
Block	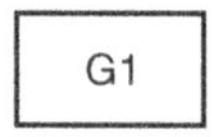	Rectangle from Shapes menu.
Connectors or arrows		Connector Elbow Arrow from Shapes menu.

4.3 Methods of Creation a Block Diagram

The user can follow either way to create a block diagram. However, the first one is highly recommended because it keeps the block as one object, and it doesn't need to group the blocks.

4.3.1 Canvas Method

Go to Insert tap, then on the Shapes menu, click on New Drawing Canvas. In this case, you will get the canvas area filled in white color as shown in Fig. 4.1-a. You can change the filling color to any color you want, see Fig. 4.1-b. Choose no fill for the canvas area to show the gridlines as shown in Fig. 4.1-c. The gridlines are important to align the blocks of the block diagram.

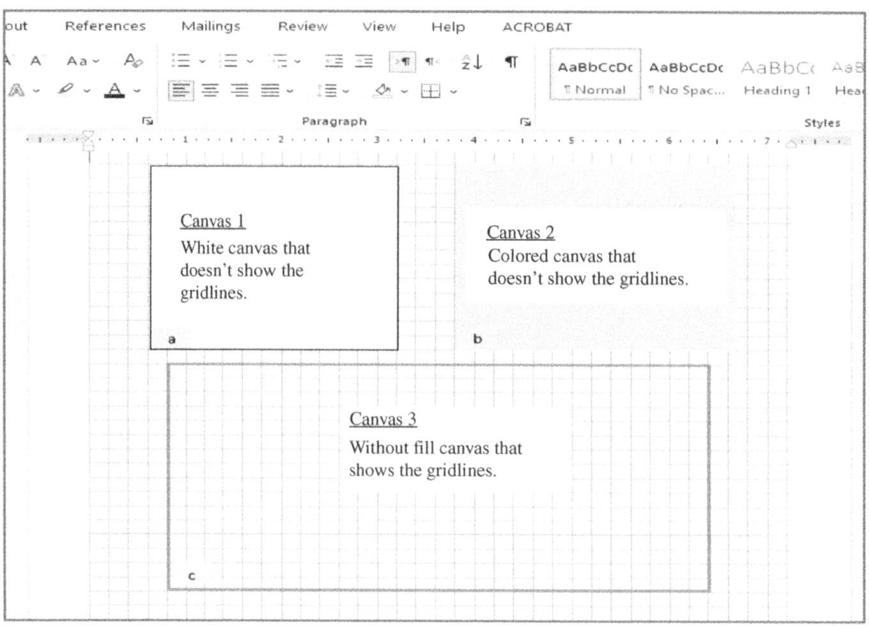

Figure 4.1 Three types of canvas.

To show the gridlines, select the canvas, then click on the Shapes Format tab. In the Arrange group, go to the Align Object icon, then open its menu. At the bottom of this menu, select View Gridlines as shown in Fig. 4.2.

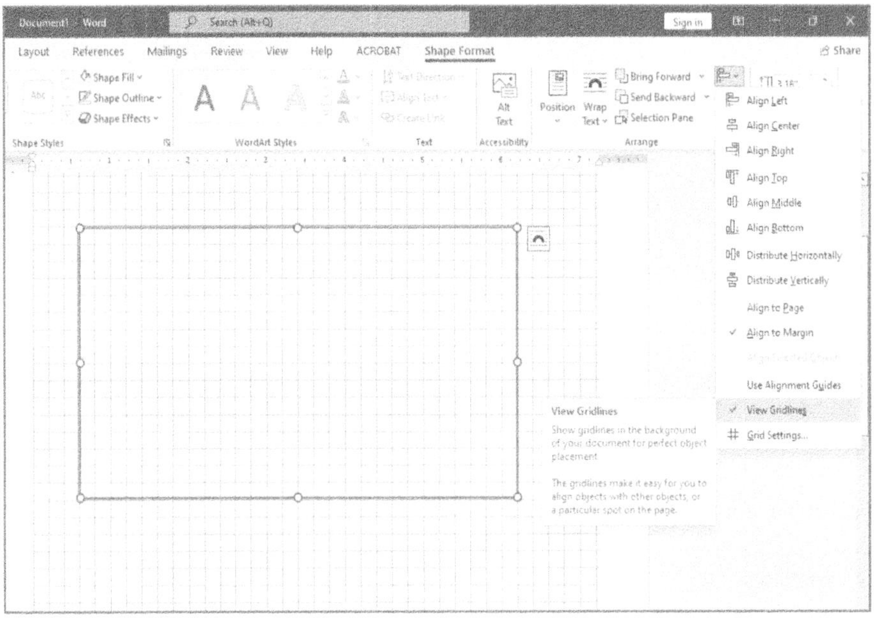

Figure 4.2 The location of the gridlines option.

4.3.2 Blank Page Method

Open Microsoft Word. Go to New, then select a Blank document. After that, start drawing your block diagram as shown in the coming examples.

4.4 Alignment Methods

4.4.1 Manual Method

Drag and drop the blocks to align them horizontally or vertically by using the mouse. Or, you can select the block and move it to the right and left or move it up and down by using the keyboard arrows (→ ← ↑ ↓). Note that you can align the blocks on a document with gridlines or without gridlines, but a document with gridlines is better and more accurate.

4.4.2 Options Method

To Align Blocks Horizontally:

- Select all the blocks that you want to align.
- On the Shape Format tab, go to Arrange group.
- Go to the Align Object icon and open its menu. Be sure that Align Selected Object is marked.
- Click on the Align Middle option. You will see your blocks are arranged horizontally. As shown in Fig. 4.3.

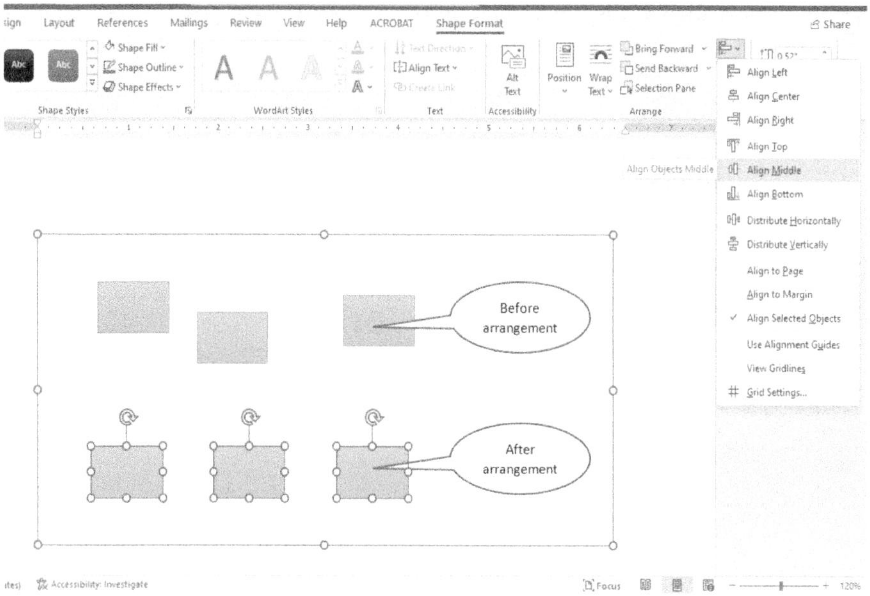

Figure 4.3 Horizontal alignment.

To Align the Blocks Vertically:

- Select all the blocks that you want to align.
- On the Shape Format tab, go to Arrange group.
- Go to the Align Object icon and open its menu. Be sure that Align Selected Object is marked.
- Click on the Align Center option. You will see your blocks are arranged vertically. As shown in Fig. 4.4.

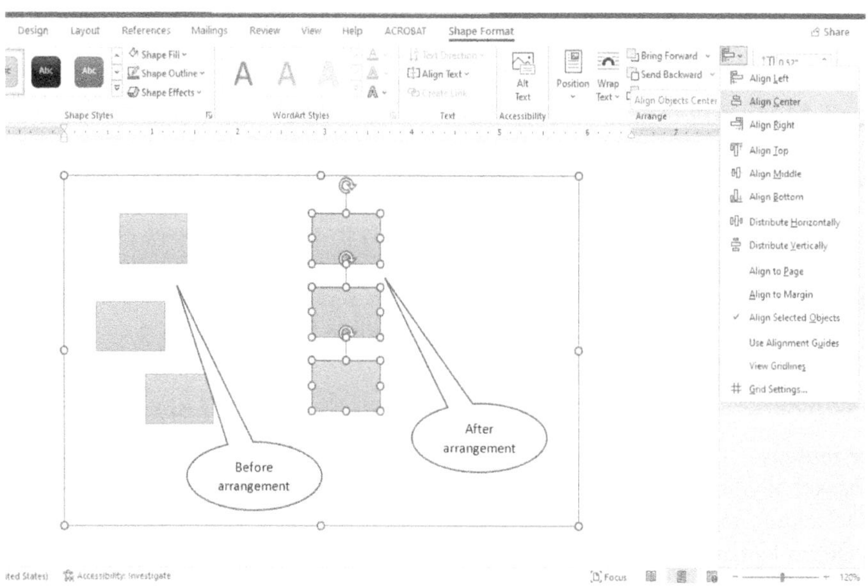

Figure 4.4 Vertical alignment.

4.5 Examples of Drawing a Block Diagram in Microsoft® Word

Here, we'll look at several examples of how to plot reduction steps on a Microsoft Word page.

Example 4.1 Draw the block diagram of Fig. 4.5 in Microsoft Word.

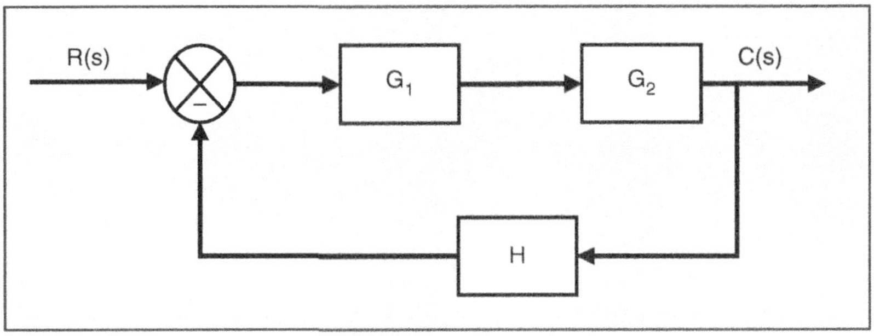

Figure 4.5 The block diagram of Example 4.1.

Procedure:

1- Open a New document in Microsoft Word.
2- Click on the Insert tab, then click on New Drawing Canvas as shown in Fig. 4.6.
3- From the Shapes menu, click on the Rectangle, then draw three rectangles. Adjust them as shown in Fig. 4.7. You can draw one rectangle, copy it, and then paste it twice to get three rectangles in total.
   Adjusting a shape means placing it in the required format, including the size, filling color, outline color, and type.
4- From the Shapes menu, go to the Flowchart and click on Summing Junction to draw the circle of the summing point.
5- Click on each rectangle to write inside it. Write G_1, G_2, and H as shown in Fig. 4.8.

Figure 4.6 Opening a new drawing canvas.

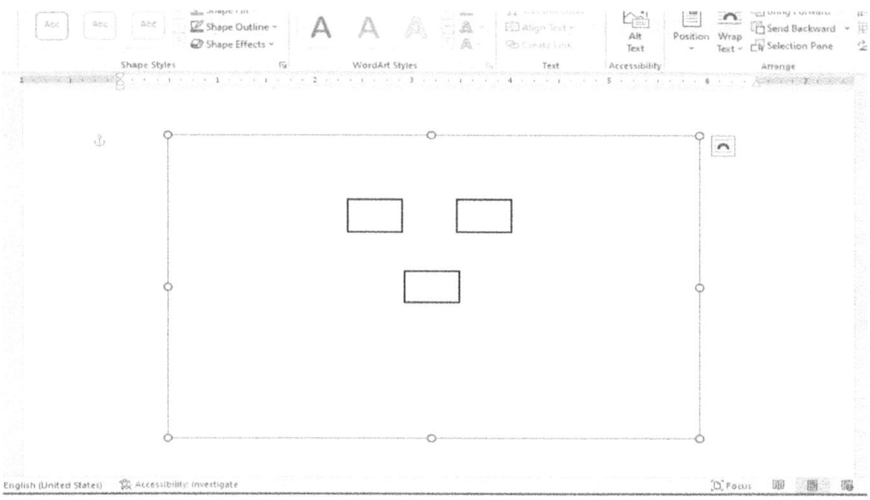

Figure 4.7 Drawing rectangles.

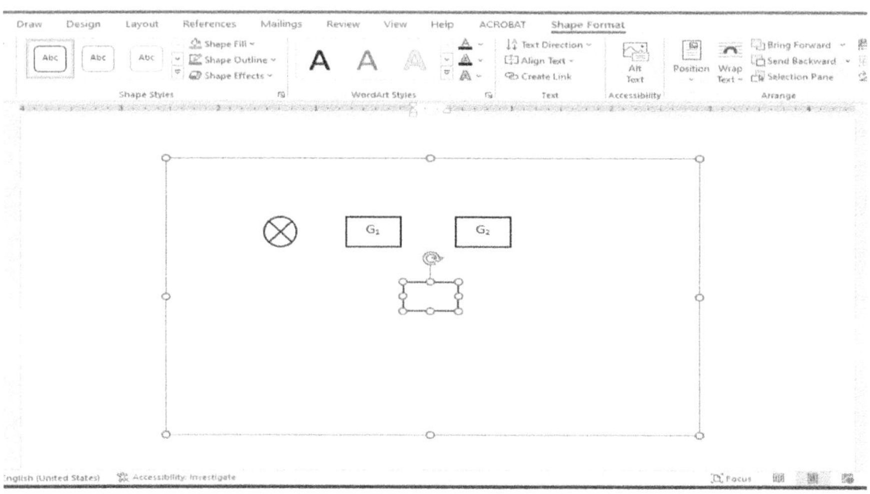

Figure 4.8 Labeling the rectangles.

6- From the Shapes menu, click on Connector to draw the lines that connect the blocks. When you connect a line to a block, you will see a green circle at the two ends of the connector line. This means the line and the block are accurately connected as shown in Fig. 4.9.
7- From the Shape menu, draw a Rectangle and then adjust it to no fill and no outline. Write R in it. Copy and paste this rectangle, then write C instead of R. Paste the rectangle one more time and write – minus sign and place it near the summing point to indicate the negative feedback as shown in Fig. 4.10.
8- Adjust the canvas size by clicking on any small circle that lies on the canvas border. Fill the canvas in blue color as shown in Fig. 4.11.

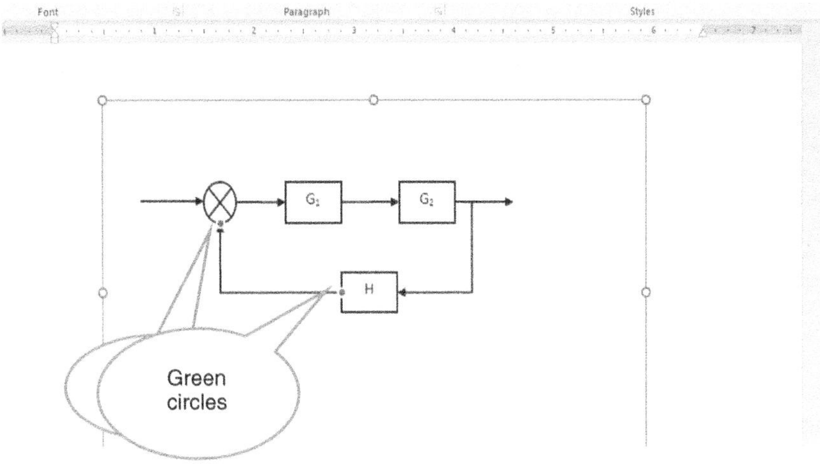

Figure 4.9 Connectors and the green circles.

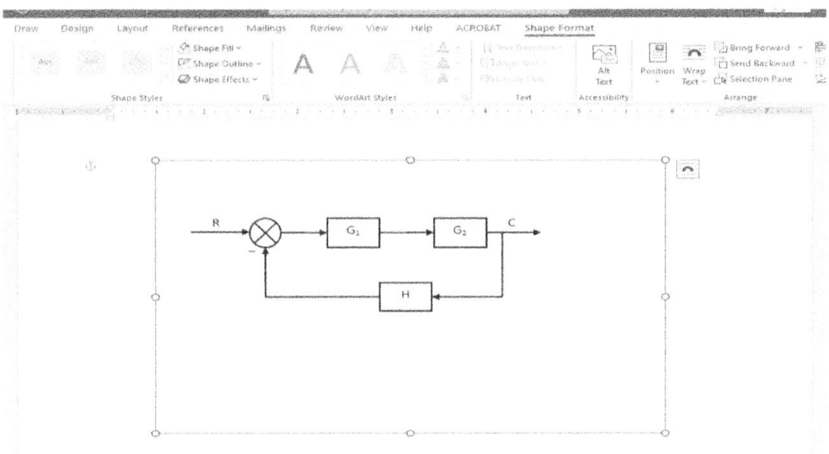

Figure 4.10 The final drawing of the block diagram.

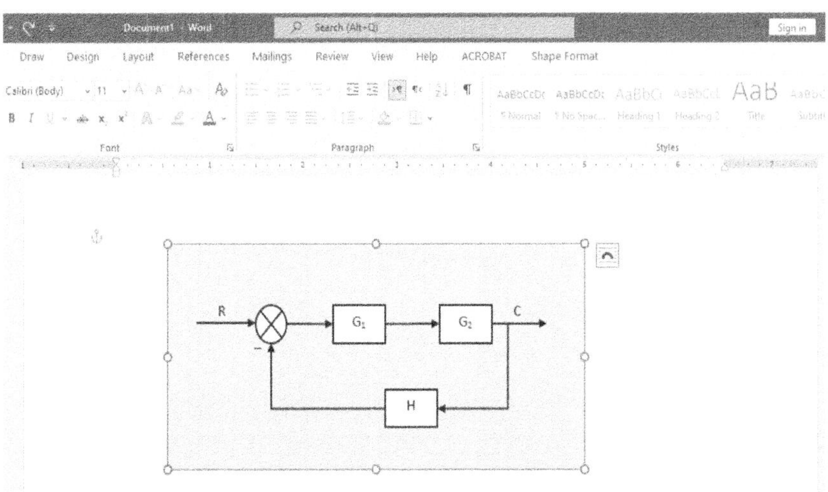

Figure 4.11 Filling the canvas background.

Notes:

- By using this procedure, you don't need to group the blocks because you drew them in a canvas area.
- It is easy to add a block or summing point to this diagram.
- By selecting a block and moving it anywhere in the canvas, the lines will move with it because they are connected to the block.

Example 4.2 Draw the block diagram of Fig. 3.11/Chapter 3 in Microsoft Word.

Procedure:

1- Open a New document in Microsoft Word.
2- Click on the Insert tab, then open the Shapes menu. Click on Insert New Drawing Canvas.
3- On the Shape Format tab, go to Arrange group. Under Align Object menu, click on View Gridlines.
4- Now your canvas is white and you can't see the gridlines. All you need to do is make the canvas area transparent (no fill) with a red border to recognize the canvas area from the whole Word page.
   - Select your canvas border, open the Shape Fill menu, and click No Fill.
   - Click on your canvas border, open the Shape Outline menu, and click on the red color.
   - Your screen should look like Fig. 4.12.
5- From the Shapes menu, click on Rectangle. Draw four rectangles and adjust them in the required format.

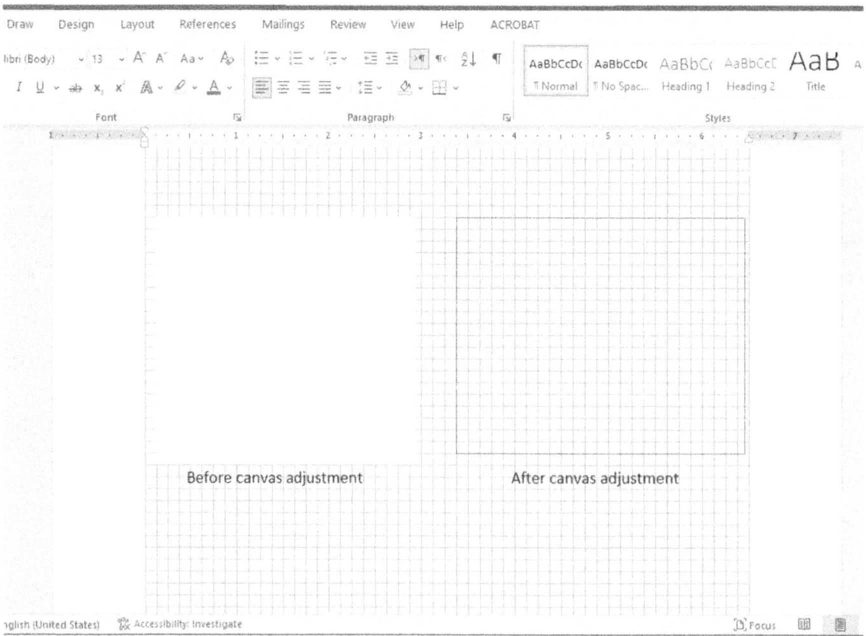

Figure 4.12 Canvas adjustments.

6- From the Shapes menu, click on Summing Junction. Draw three summing points and adjust them in the required format. Click on each rectangle to write inside it. Write G_1, G_2, H_1, and H_2 as shown in Fig. 4.13.

7- Align G_2, H_1, and H_2 blocks vertically by using the mouse and follow the gridlines to adjust the position for each block as shown in Fig. 4.14.

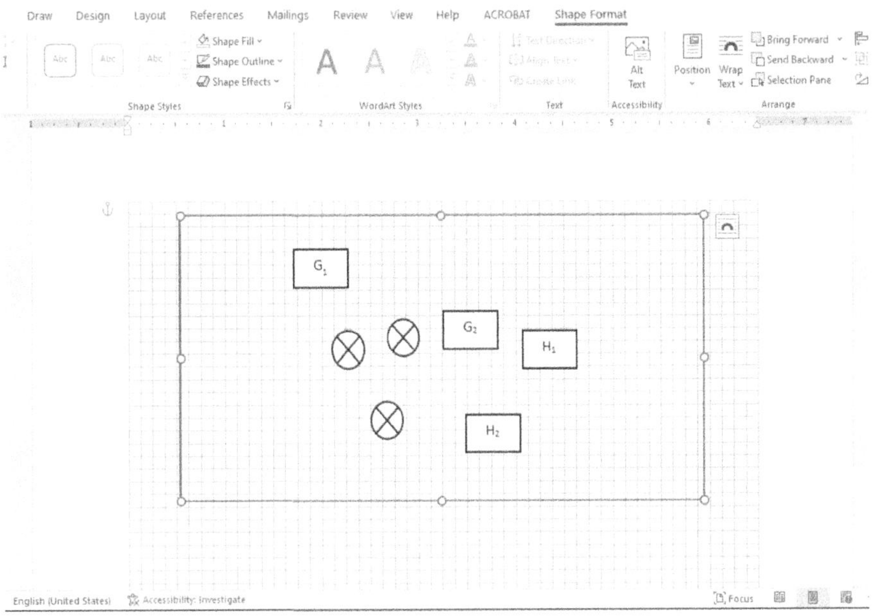

Figure 4.13 Labeling blocks.

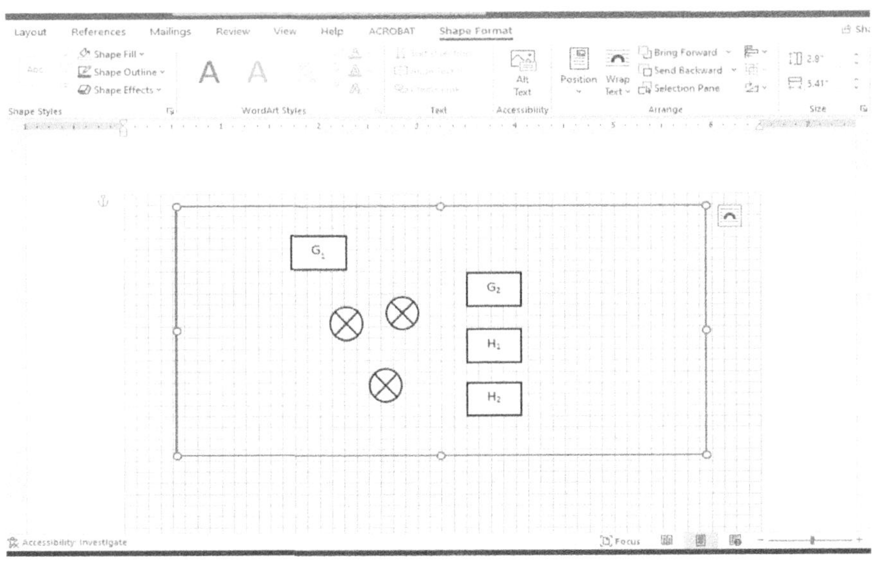

Figure 4.14 Blocks alignment by using the mouse.

8- Align the first and second summing points horizontally, then align the second and the third summing points vertically. Place block G_1 above the first summing point as shown in Fig. 4.15.
9- From the Shapes, click on Connector to draw lines that connect the blocks and the summing points with each other.
10- Write R at the beginning and C at the end of the block diagram. Write plus and minus in the summing points in separate text boxes. Then, after labeling the required parts, adjust your canvas border as shown in Fig. 4.16.
11- Now fill your canvas background in a desired color and turn off the gridlines as shown in Fig. 4.17.

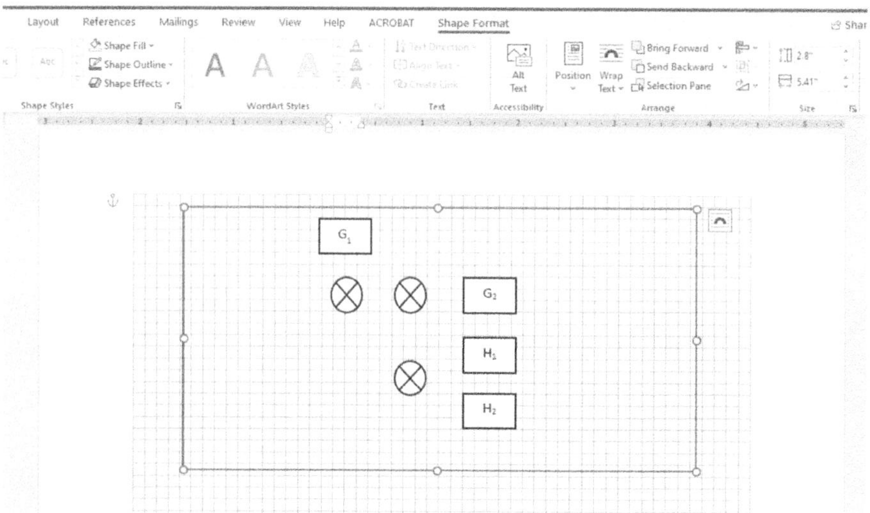

Figure 4.15 Summing points alignments.

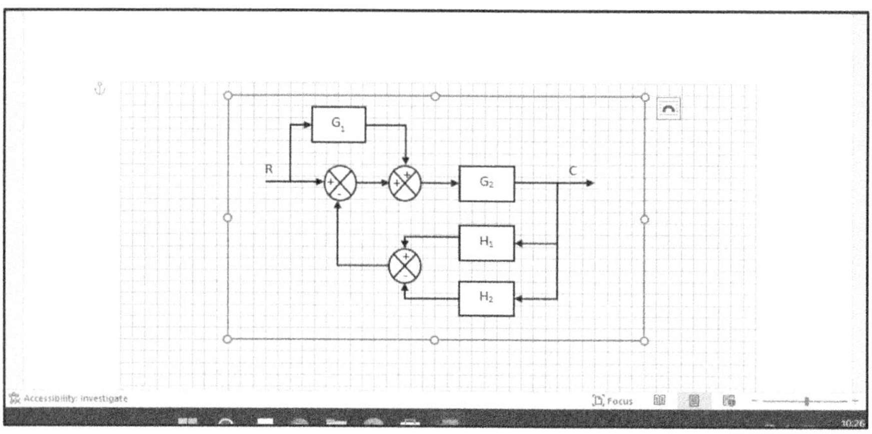

Figure 4.16 Connecting the blocks and complete labeling.

Note:

If you want to fill any block with any color, you can do that easily. Just click on or select the block, then on Shape Format, go to Shape Fill and choose a color from the menu.

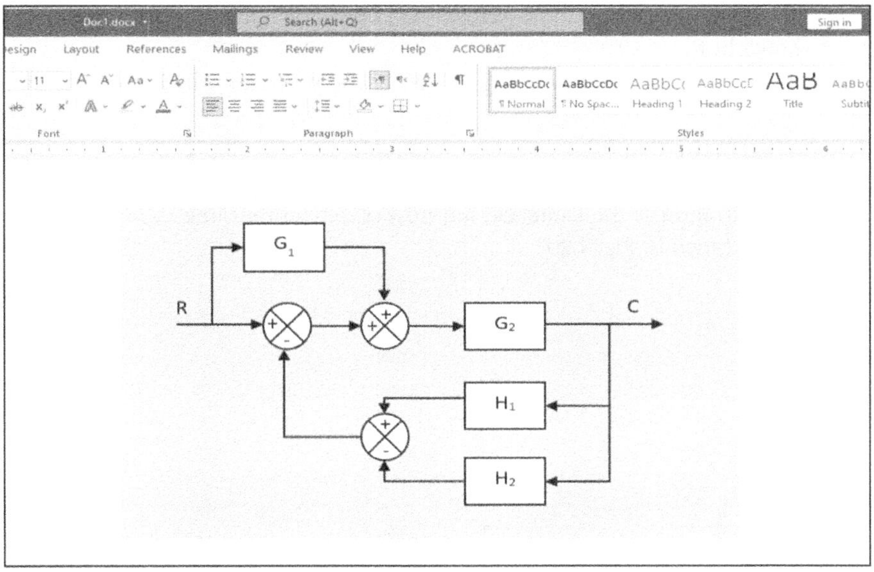

Figure 4.17 Filling the canvas background and finishing the block diagram.

Example 4.3 Draw the block diagram of Fig. 3.1-b'/Chapter 3 in Microsoft Word.

Procedure:

1- Open a New blank document in Microsoft Word.
2- Click on the Insert tab, then from the Shapes menu, click on Rectangle. Draw five rectangles, then adjust and label them as required.
3- Go to the Shapes menu and click on Summing Junction. Draw two summing points.
4- Align the blocks and the summing points as required by using the mouse or the keyboard. Your figure should look like Fig. 4.18.

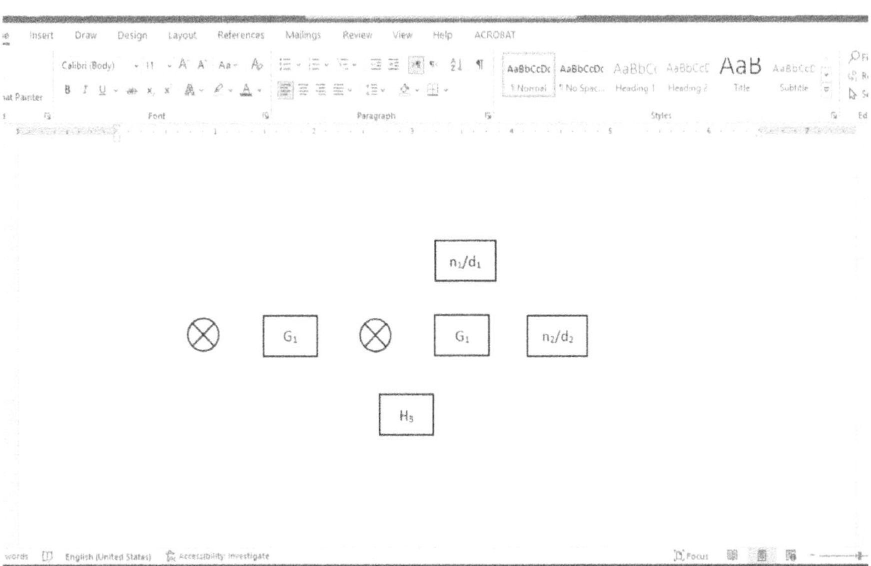

Figure 4.18 Drawing and alignments of the required blocks and summing points.

5- Go to Shapes and click on Connectors: Elbow Arrow, then draw and adjust the lines that connect the blocks as shown in Fig. 4.19.
6- Label input and output R and C in small text boxes. Write the plus and minus signs on the summing point.
7- Now for illustrations. From the Shapes menu, click on Oval to draw an ellipse around the two series blocks. Color the ellipse line and write "series" in a text box.
8- Repeat step 7 for the illustration of the feedback, but now select another line color and write feedback in a text box as shown in Fig. 4.20.

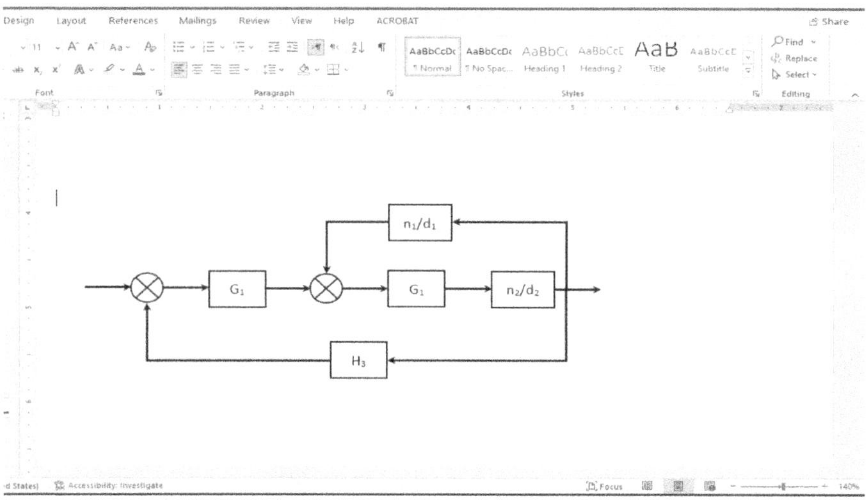

Figure 4.19 Connecting the blocks.

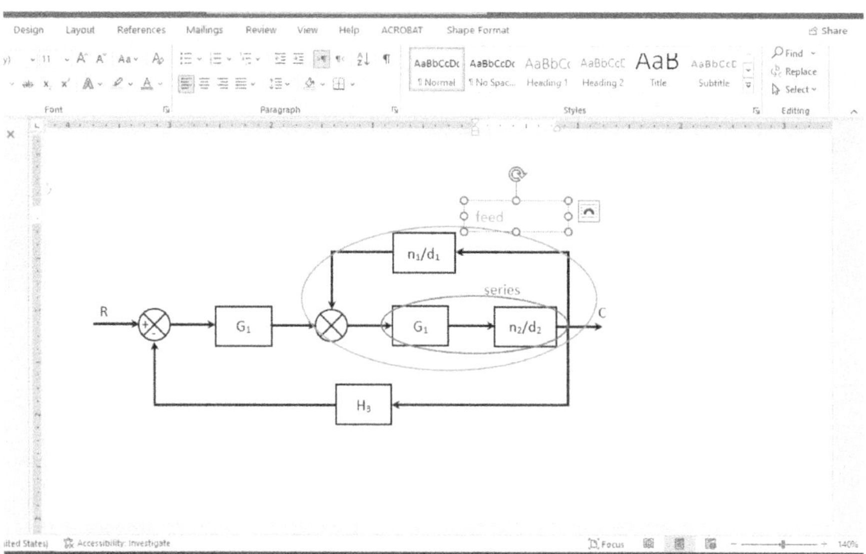

Figure 4.20 The block diagram after completing steps 6 to 8.

9- After the completion of this block diagram, click the Home tab, go to Select, and click on Select Object. Select all the block diagram parts and group them all as shown in Fig. 4.21.

10- Your figure should look like Fig. 4.22.

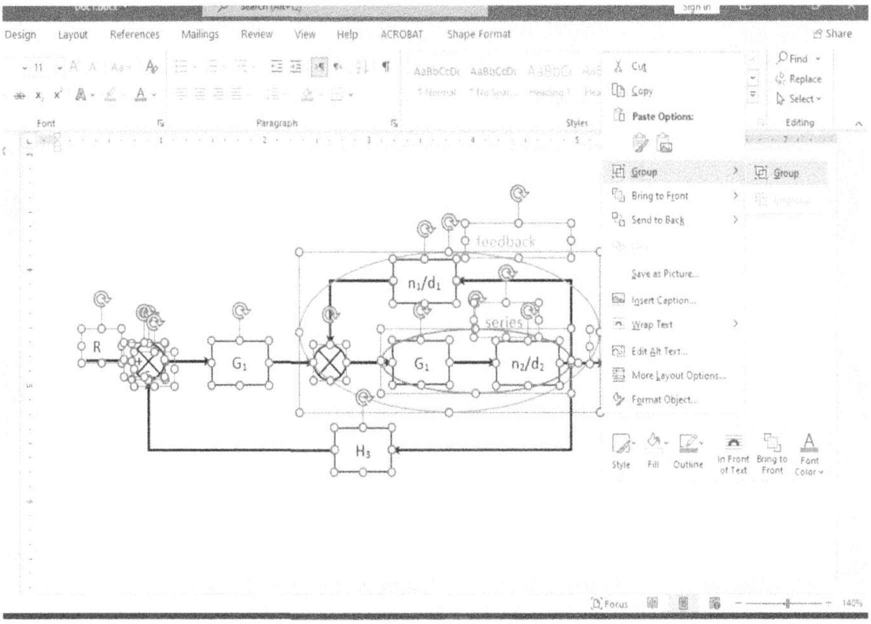

Figure 4.21 Using the Group option.

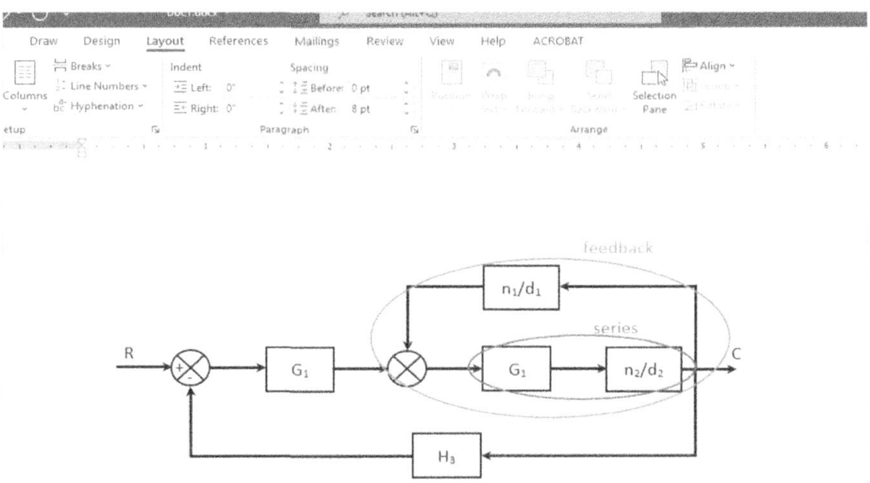

Figure 4.22 Finishing the block diagram.

11- To set a background to your figure, go to the Shapes menu and click on Rectangle. Draw a rectangle and fill it in any color. Then, on the Shapes Format tab, open the Send Backward menu and click on the Sent to back option. Now the colored rectangle is behind your block diagram as shown in Fig. 4.23.

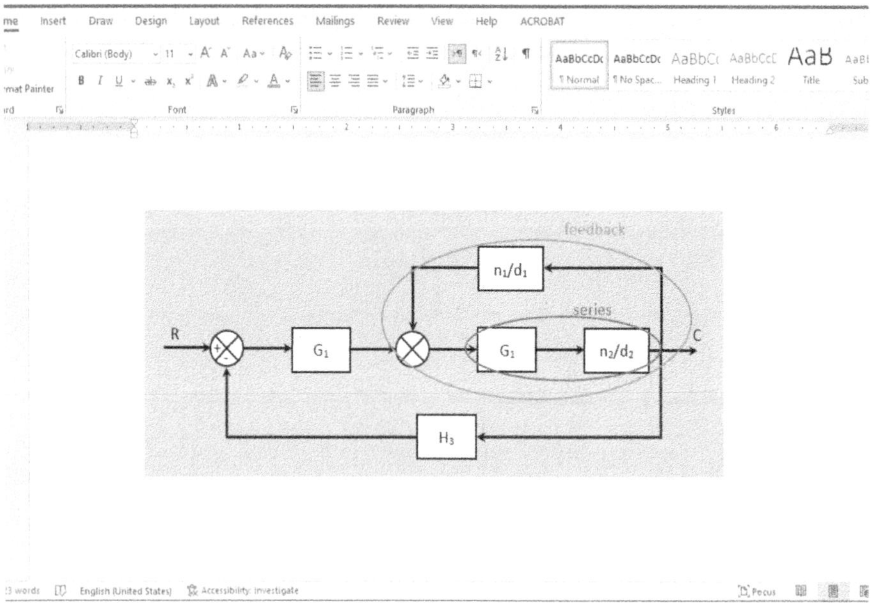

Figure 4.23 The final block diagram.

Notes:

- By using this method, you need to group all the block diagrams' parts.
- If you have difficulty in selecting object(s) in Word, follow this note. Click the Layout tab, then click on Selection Pane. A Selection menu will appear on the right side of your computer screen as shown in Fig. 4.24. By clicking on any object in the menu you will see the appearance of small circles around the selected object.

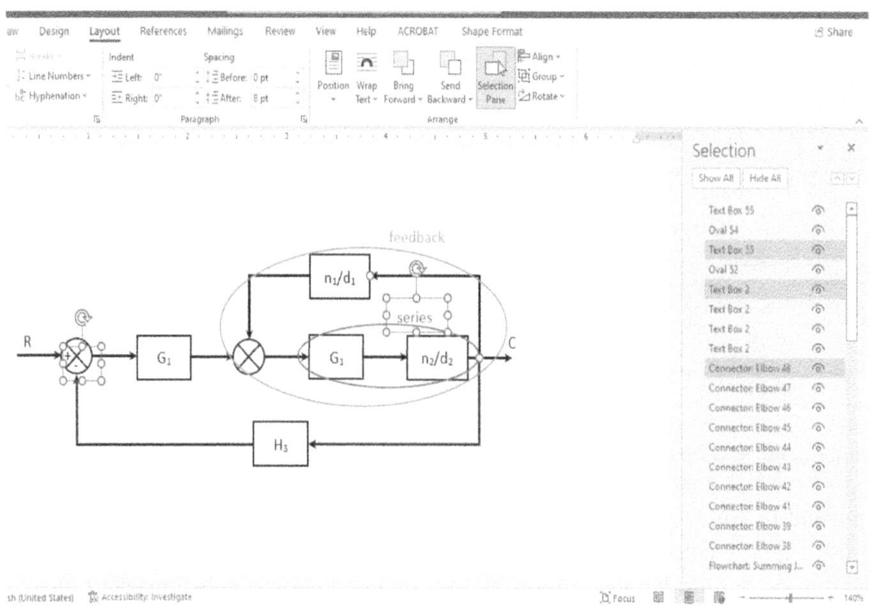

Figure 4.24 Selection Pane screen.

Example 4.4 Draw the block diagram of Fig. 3.4/Chapter 3 in Microsoft Word.

Procedure:

1- Open a blank document in Microsoft Word.
2- To turn on gridlines: Click on the View tab and then mark the Gridlines.
3- Draw three summing points and three rectangles, then label them as required as shown in Fig. 4.25.
4- To align these six objects: Select all these objects. Then, on the Shapes Format tab, open the Align Object menu and click on Align Middle. After that, open the same menu one more time and click on Distribute Horizontally so your figure will look like Fig. 4.26. Note that if you exceed the edges of the grid, use the keyboard arrows to move these blocks in any required direction.

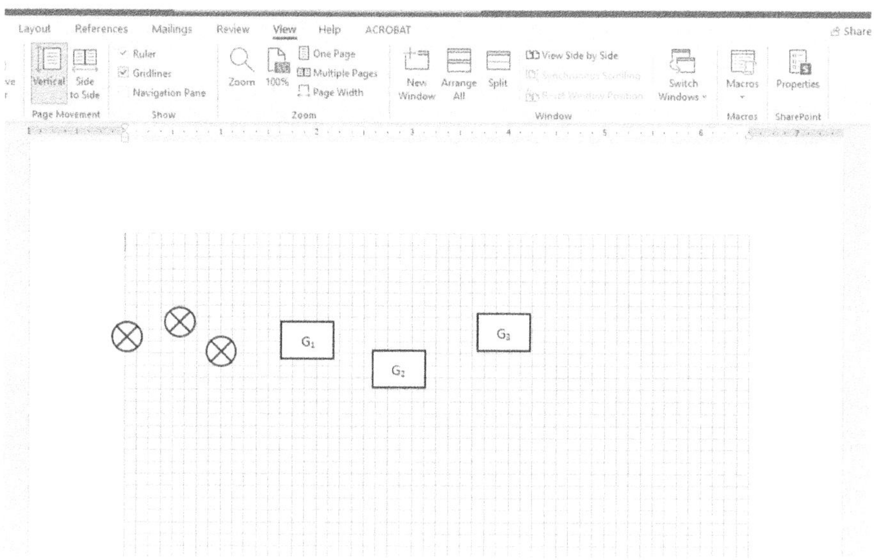

Figure 4.25 The block diagram after completing steps 1 to 3.

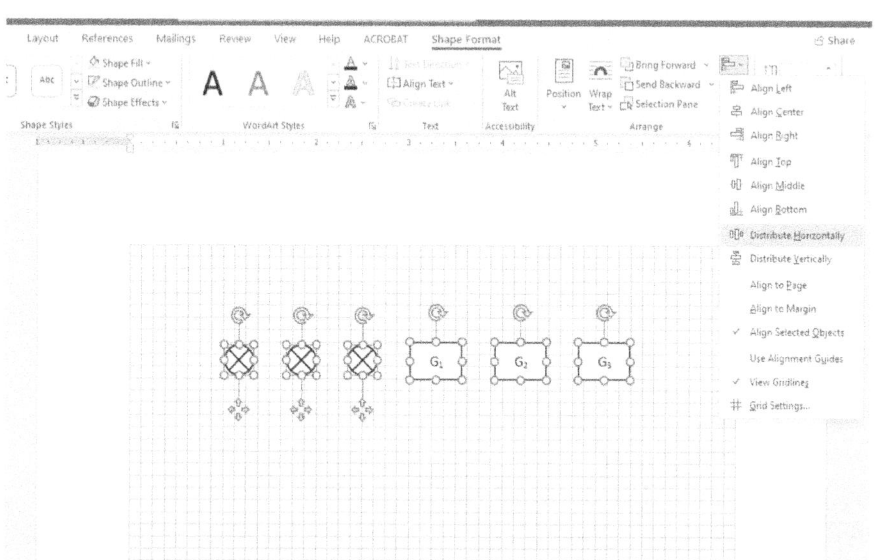

Figure 4.26 The blocks and summing points alignments.

5- To draw the H's blocks: select the three G's blocks, copy and paste them. Change their labels to H_1, H_2, and H_3 as shown in Fig. 4.27.

6- To align a group of blocks vertically: select blocks G_1, H_1, H_2, and H_3. Click on the Shapes Format tab, then go to the Align Object and open its menu. Click on the Align Center option. After that, open the same menu one more time and click on Distribute vertically. Use the keyboard arrows to adjust the blocks in the required position as shown in Fig. 4.28.

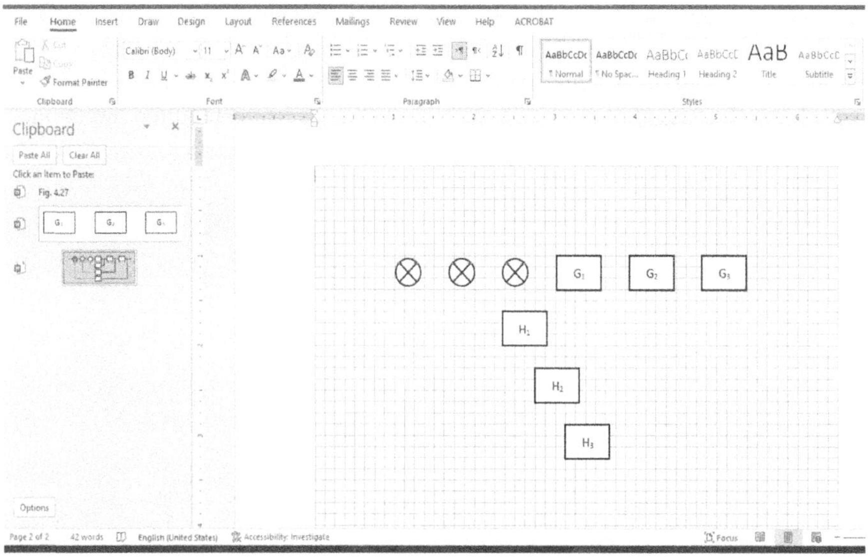

Figure 4.27 Drawing the H's blocks.

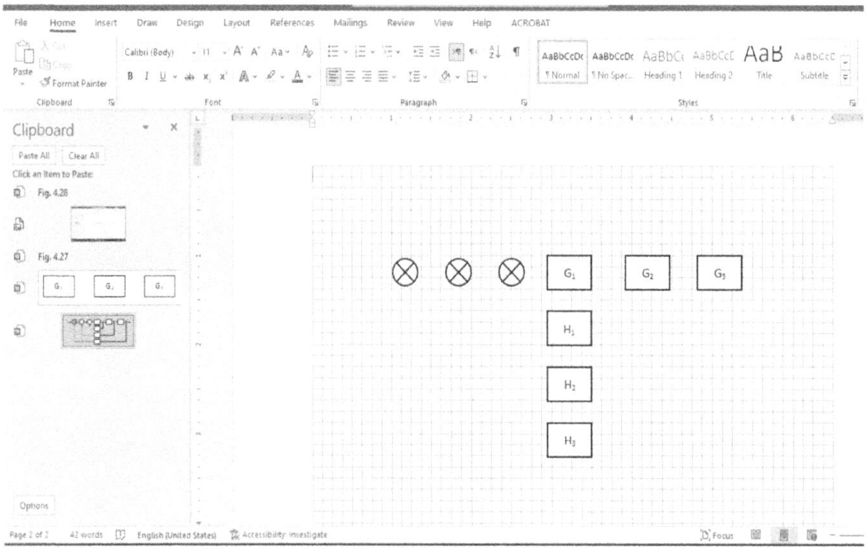

Figure 4.28 The H's block alignment.

7- From the Shapes menu, use the Connectors to draw the lines to connect the blocks. Label the input and output, and place plus or minus signs on the summing points. Your figure should look like Fig. 4.29.

8- Select all objects and group them to be a single object. Then draw a rectangle, fill it in a desired color, and then send it behind your diagram. Finally, select the rectangle and the diagram and group them. Click the View tab, then unmark the Gridlines to turn them off. Your figure should look like Fig. 4.30.

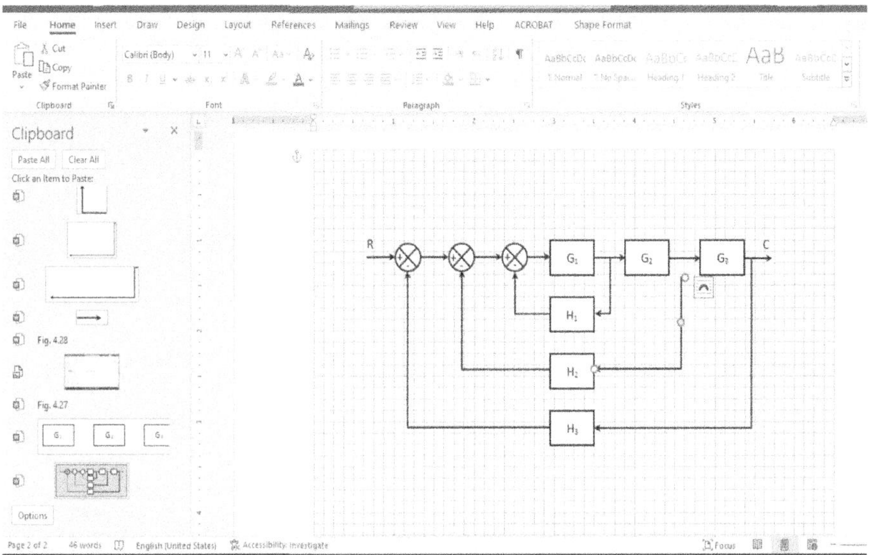

Figure 4.29 Connecting the blocks with connectors.

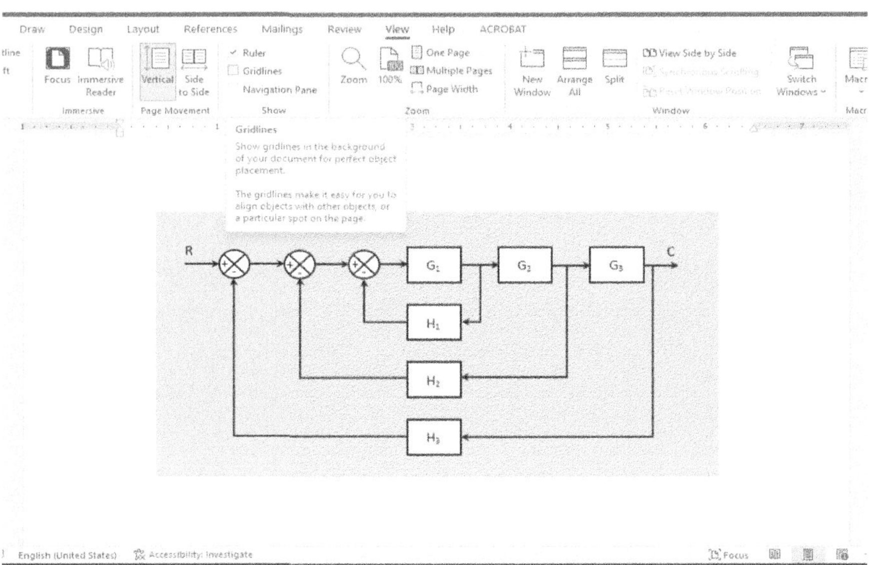

Figure 4.30 The final block diagram.

Note:

- To make sure that all objects in your block diagram become one group, select the group and press the Delete key on the keyboard to see if there are more objects that aren't in the group. If so, go to the Layout tab, then open Selection Pane menu. Select any ungrouped object/s by clicking on the object/s from the menu, then add them to your group.
- The user can use any of these recommended methods to draw any block diagram.

For More Practice:

Draw any block diagram using the two discussed methods.

Review Questions

1 Draw 4 blocks.

2 Align them horizontally.

3 Align them vertically.

4 Draw the block diagram of Fig. 3.5/Chapter 3.

5 Draw the block diagram of Fig. 3.6/Chapter 3.

Note:

You can build any block diagram following the previously described methods or create a new one in Microsoft Word. If you lack knowledge in Microsoft Word or you prefer to use other software like Viso or LaTex, you can do so. Just take the notion from this chapter and apply it to your software. Each software has specific features.

5

Block Diagram Presentation in Microsoft® PowerPoint

5.1 Introduction

Presentation is required for teaching, attending a conference, or planning a seminar. The most critical consideration is for whom the presentation is being created. Readers may struggle to link the objectives of the presentation with its audience. So, this chapter teaches how to make a well-organized and well-structured presentation for block diagram reduction steps by using Microsoft PowerPoint. First, make the reduction steps and diagrams on a piece of paper. Next, decide which diagram you want to show and how. Finally, follow the presentation methods that are presented here, and you can also create another one.

Learning Objectives:

- To learn how to present a block diagram in Microsoft PowerPoint.
- To learn which presentation method to use for a specific block diagram.
- To learn how to organize the reduction steps in a PowerPoint slide.
- To learn how to import the MATLAB script file into the presentation file.
- To learn the best word used to move to another slide.

5.2 Organizing a Slide

This chapter demonstrates various slide organization approaches. It is determined by a variety of elements, including the objective of the presentation, the topics covered, and the audience for whom the presentation is being prepared. You can either use an existing approach or create one.

The First Method

This method needs one slide and several animations.

1- Open a new blank presentation file in Microsoft PowerPoint.
2- Divide the slide into four sections as shown in Fig. 5.1. In the first section, place the MATLAB code without any animations. In the second section, place the original block diagram without

Block Diagram Implementation for Control Systems: Including MATLAB® Programming, Microsoft® Word and PowerPoint Presentation, First Edition. Ruba Al-Mulla Hummadi.

any animations to keep returning to it while explaining the steps or stages of reduction. Moving to the third section, place the reduced block diagrams for each stage in cascade form, starting from the first stage until the last one. Here, it is important to use animations, but use one type only to avoid confusing the audience. Use Exit animation, for example. The idea is just like when you have a book and you are turning the pages after finishing explaining each page. Lastly, in the fourth section, place the overall transfer function of the system or place the transfer function for the current step as shown in Example 5.1.

The Second Method

This method is done with more than one slide and without any animations.

Arrange each slide as shown in Fig. 5.1 and place one figure in part 3 only. In other words, for the first slide, arrange it as in Fig. 5.1 and place the first reduction step in part 3. Arrange the second slide as in Fig. 5.1 and place the second reduction step in part 3. Repeat until finishing all the reduction steps. In part 4, write the transfer function for each step, or leave it blank, or write just the final transfer function. When you start your slide show, you will have only one step of reduction on each slide as shown in Example 5.2.

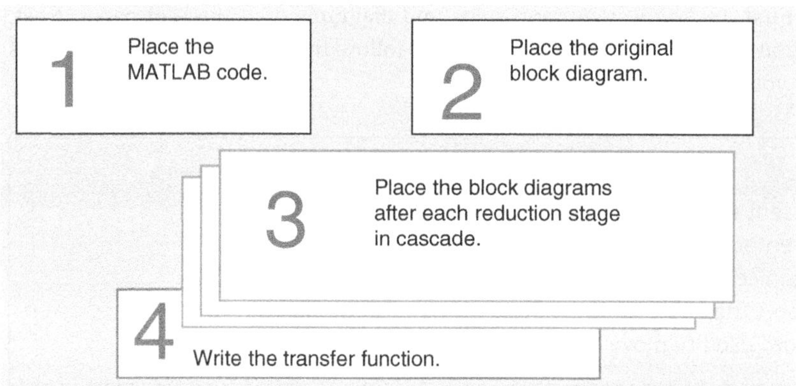

Figure 5.1 Organizing a PowerPoint slide for block diagram reduction steps.

The Third Method

Some block diagrams are simple, neat, and their reduction steps are repeatable. You can prepare a simple and quick presentation by annotating your block diagram and animating these annotations as it will be shown in Example 5.3.

The Fourth Method

Organize your slides as illustrated in Figs. 5.2 and 5.3.

1- Organize the first slide as in Fig. 5.2, then:
   - Place the MATLAB code in part 1.
   - Place the original block diagram in part 2.

- In part 3, place step 1 of reduction.
- In part 4, place the transfer function if you want to show it after each step. If not, leave part 4 empty.

2- Organize the second slide as in Fig. 5.3, then:
- Place the MATLAB code in part 1.
- Place step 1 of reduction in part 3
- In part 5, place step 2 of reduction.
- In part 4, place the transfer function for the current step.

3- Organize the third slide as in Fig. 5.3, then:
- In part 1, place the MATLAB code.
- Place the reduction step 2 in part 3 and the reduction step 3 in part 5.
- In part 4, place the transfer function.

4- If you need more slides, continue following the same process; each time, place the previous reduction step in part 3 and place the next reduction step in part 5 as shown in Example 5.4.

There is no need for any animations.

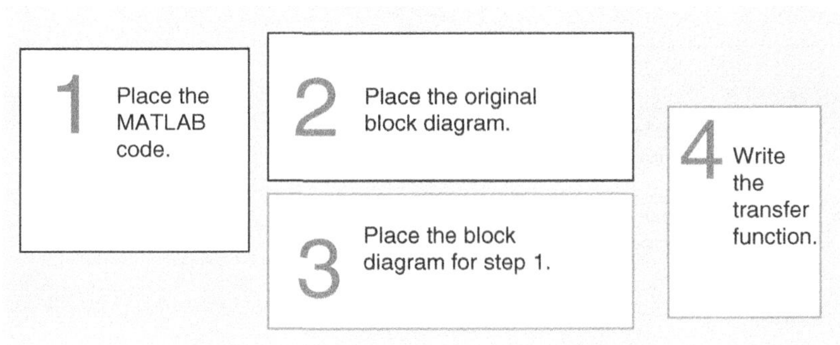

Figure 5.2 First slide arrangement.

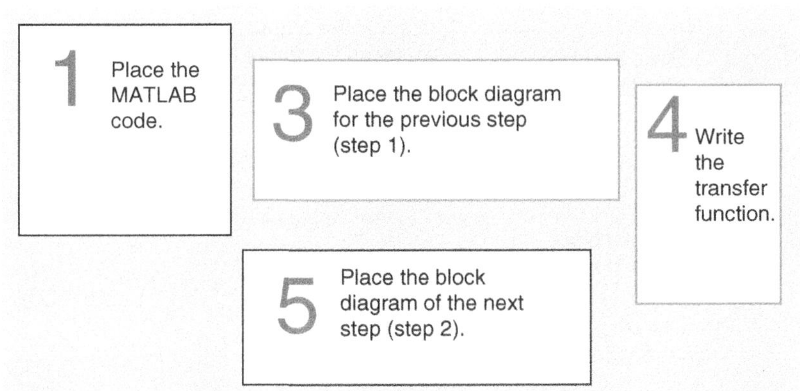

Figure 5.3 Second and other slides arrangement.

The Fifth Method:

1- Arrange the first slide as shown in Fig. 5.4, then:
   - Place the MATLAB code in part 1.
   - Place the original block diagram in part 2.
   - In part 3, place step 1 of reduction, then animate it. Move it up until it replaces the figure in part 2.
   - In part 4, place the transfer function if you want to show it after each step. If not, leave part 4 empty. Animate the transfer function if you want.
   - In part 5, place the next reduction step. Be sure to place the figure in part 5 behind the figure in part 3.

2- Arrange the second slide as shown in Fig. 5.5. Follow the same process as in the first slide, but with different reduction steps. Place step 1 in part 2, step 2 in part 3, and step 3 in part 5, then animate the figure in part 3 as it will be shown in Example 5.4.

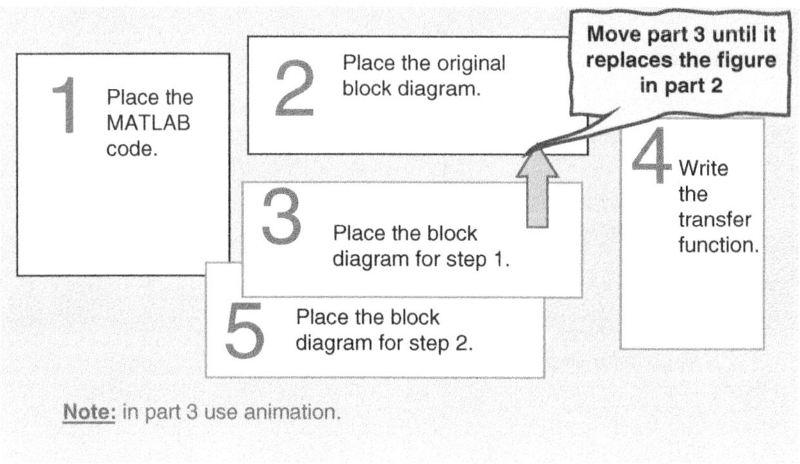

Figure 5.4 First slide organization.

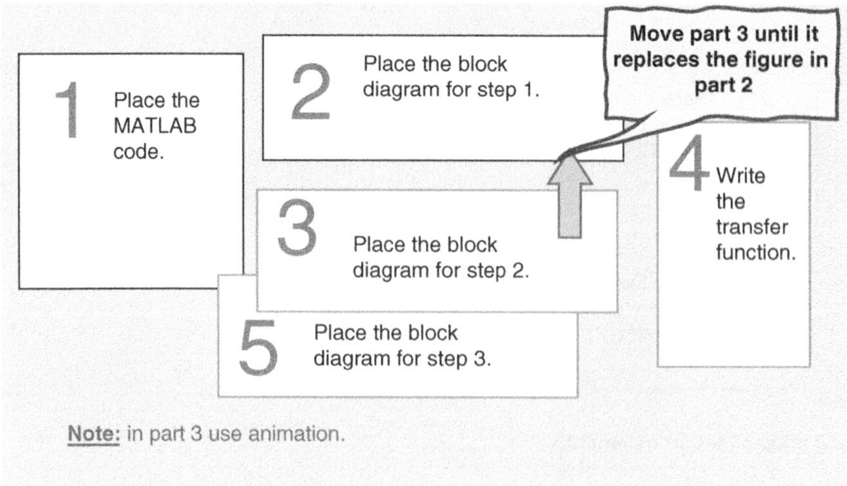

Figure 5.5 Next slide arrangement.

3- If you need more slides, do the same process as in item 2, but with different reduction steps. Place step 2 in part 2, step 3 in part 3, and step 4 in part 5, then animate the figure in part 3.

5.3 Importing a MATLAB Script into a Presentation File

Two ideas are illustrated to import the MATLAB script file into the presentation file:

Idea 1	**Idea 2**
• Open your m-file in MATLAB as shown in Fig. 5.6. • Select the program. Right-click and select copy from the menu as shown in Fig. 5.6. • Return to your PowerPoint slide and add a text box. Then, click paste. You will find your m-file in a text box in your presentation as illustrated in Fig. 5.7. It is recommended to use `Courier New` font with a size of 10 because it is the MATLAB default font. If the font size in the slide show is too small or unclear, you can adjust it as required.	• Open your m-file in MATLAB as shown in Fig. 5.8. • On keyboard, press Ctrl & PrtSc keys to get a print screen photo for your m-file. • Return to your PowerPoint slide and click paste to add a photo of the m-file in the presentation file as illustrated in Fig. 5.9. • Make any required adjustments to the photo and save the work.

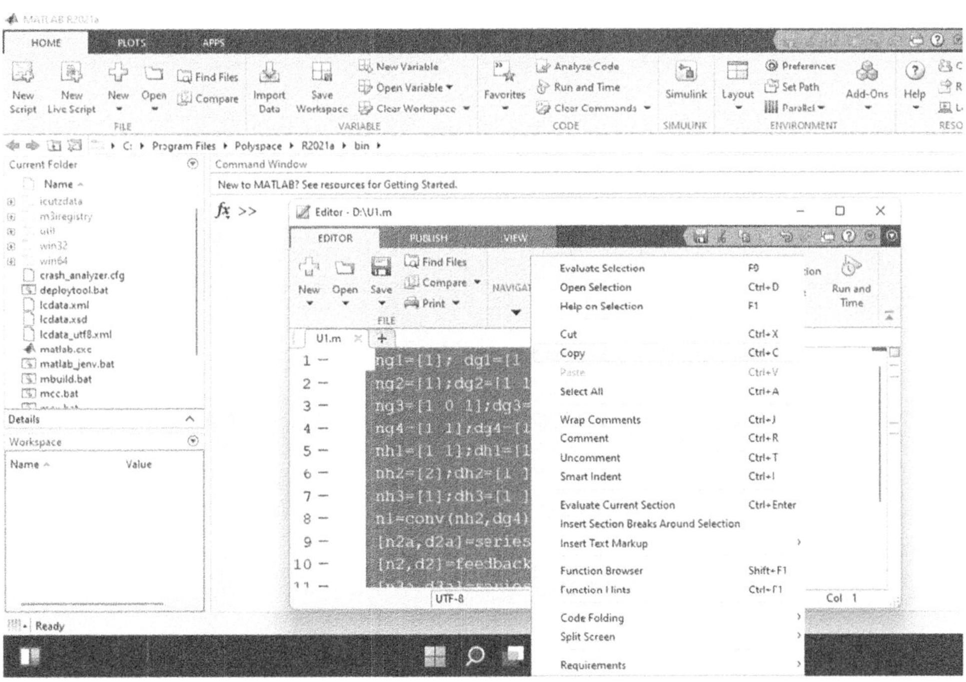

Figure 5.6 MATLAB Screen.

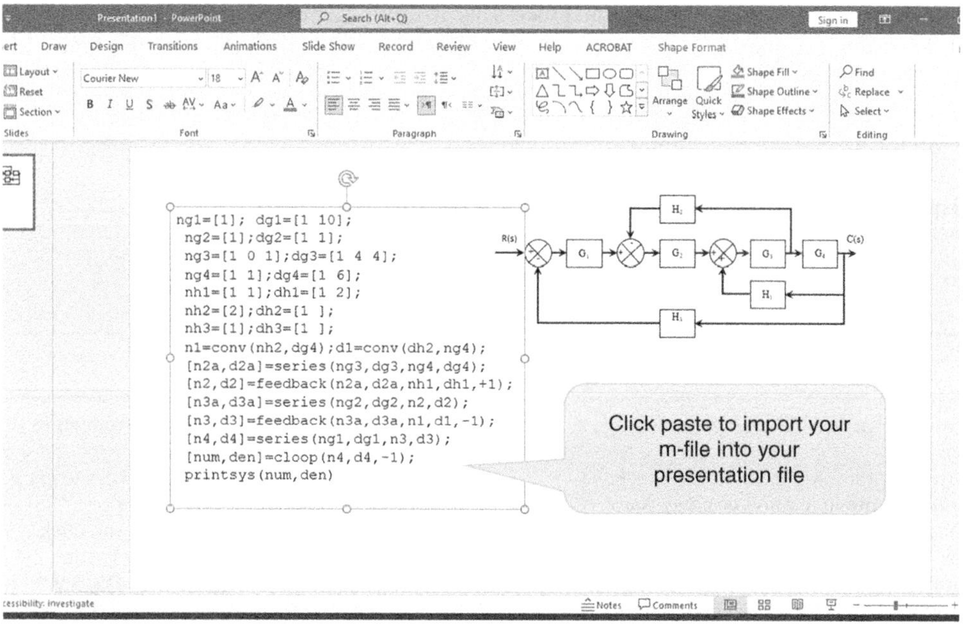

Figure 5.7 PowerPoint Screen.

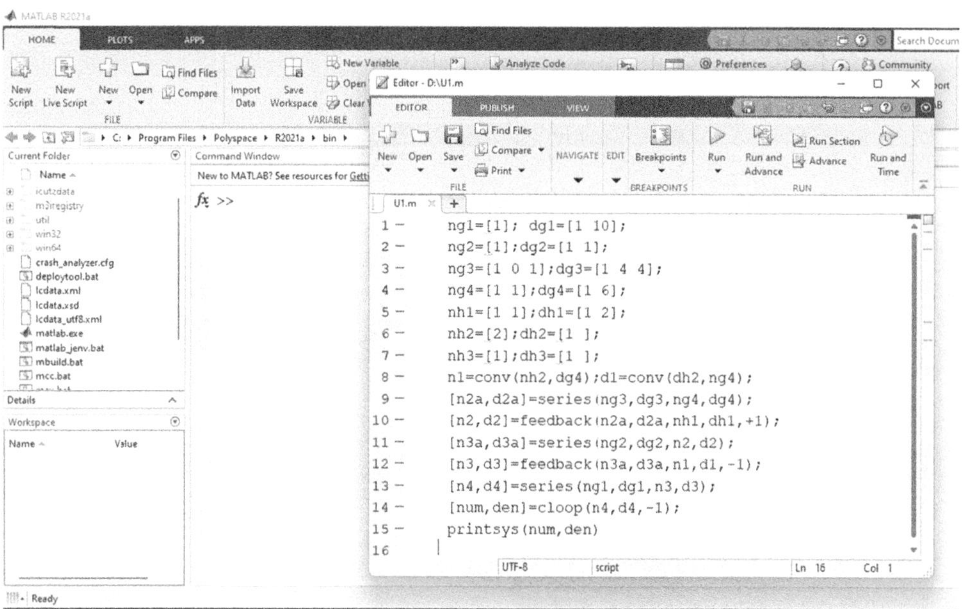

Figure 5.8 MATLAB Screen and m-file window.

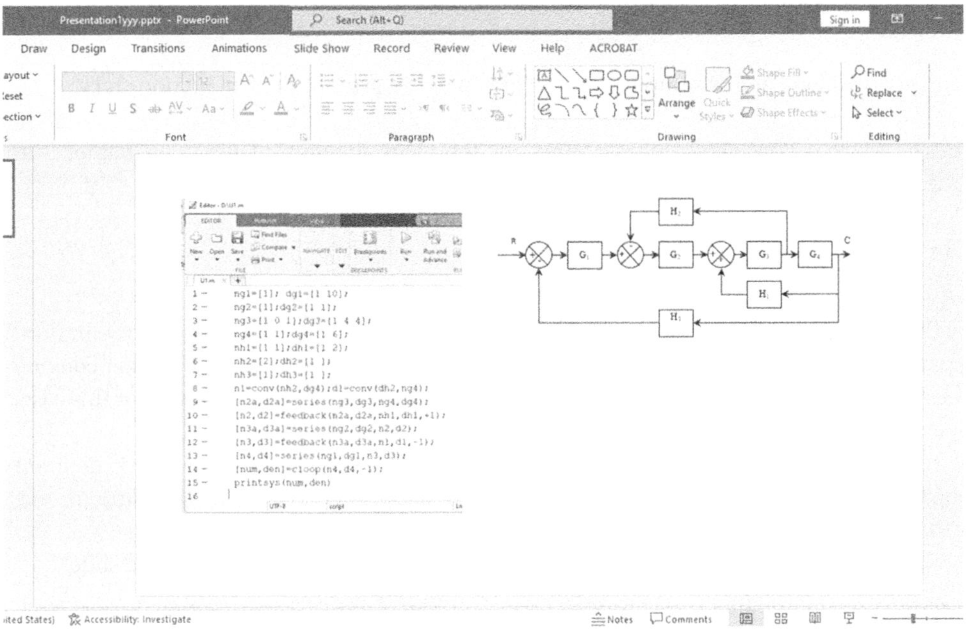

Figure 5.9 PowerPoint Screen with m-file photo.

5.4 Expressions Used to Turn to Another Slide

To switch between slides, use the following expressions:

- Starting with, the first slide shows, or first: use any one of them for the first slide only.
- Second, Third,…, etc.: used for the following slides.
- Next: used for any next slide.
- Moving on: used to turn to another slide.
- Before the last or Before finish: used specifically for before the last slide.
- Lastly or Finally: used specifically for the last slide.

Note:

Don't stick with one word or expression, for example, the word "Next", because your audience will get bored easily and will count how many Nexts have you said instead of focusing on the material. You have to keep your audience connected with you all the presentation time.

5.5 Examples of Presentation

In this item, you will look at few samples and see how we may present their reduction steps.

Example 5.1 Make a PowerPoint presentation for the block diagram of Fig. 3.1/Chapter 3 to demonstrate its reduction steps.

Process: by following the first method

1- Open a New blank presentation file in Microsoft PowerPoint.
2- Follow the slide organization of Fig. 5.1 as explained above in the first method of presentation.
3- To fill part 1 of Fig. 5.1, import your MATLAB code into your presentation file by either copying the code or adding a print screen photo of your MATLAB code as explained above. For this case, we copied the code and pasted it into the presentation file as shown in Fig. 5.10.
4- To fill part 2 of Fig. 5.1, Open the Microsoft Word file that contains your block diagram and click copy. Return to your PowerPoint slide and click paste. The slide should look like Fig. 5.10.
5- To fill part 3 of Fig. 5.1 and arrange the appearance of all the reduction stages on the slide.
   - After preparing all the block diagrams as shown in Figs. 3.1-a to 3.1-d/Chapter 3, start from the last one (Fig. 3.1-d) and place it in part 3 of Fig. 5.1 without any animations as shown in Fig. 5.11.
   - From the Shapes menu, select Rectangle, then draw a rectangle and fill it with any light color. Go to Microsoft Word and copy Fig. 3.1-c, then return to PowerPoint and paste it inside the colored rectangle as shown in Fig. 5.11.
   - Repeat the previous step for Figs. 3.1-b and 3.1-a as shown in Fig. 5.12.
   - Now make any suitable adjustment, for example, decrease the MATLAB code font size, adjust the colored rectangles with their figures, and then group each rectangle with its figure. After finishing, place Figs. 3.1-d to 3.1-a in cascade. Set the slide background to any desired color as shown in Fig. 5.13.
     Note: Avoid using any texture or photo in the slide background to avoid confusion.
   - Now start applying the animation based on mouse click, starting from Fig. 3.1-a and ending with Fig. 3.1-c as shown in Fig. 5.14. Use Exit animation type to let the current figure flies out of the page and allow the next figure to be seen.
6- To fill part 4 of Fig. 5.1,
   Display the final transfer function on the screen as shown in Fig. 5.15.
7- On the slide show, your presentation should look like Fig. 5.16. See how the slide is changed after each mouse click, knowing that all these figures are organized in one slide. For example, after the first mouse click, the first step of reduction will fly out of the slide to let the second step be seen. Continuing on the same process allows the other steps to be shown. Then, on the last click, the transfer function will be displayed.
8- Save the file under a chosen name.

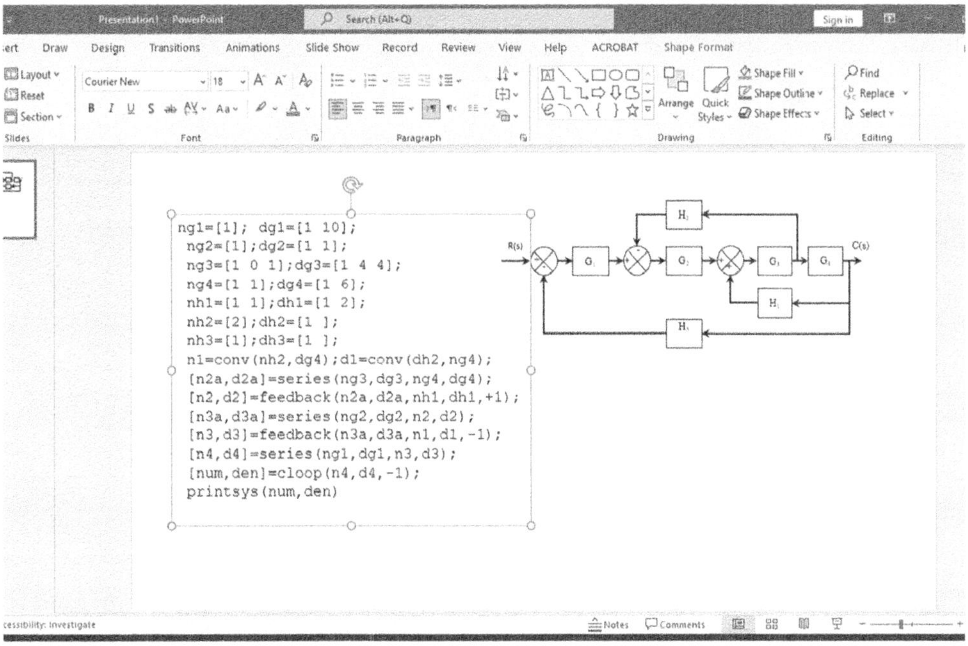

Figure 5.10 PowerPoint Screen.

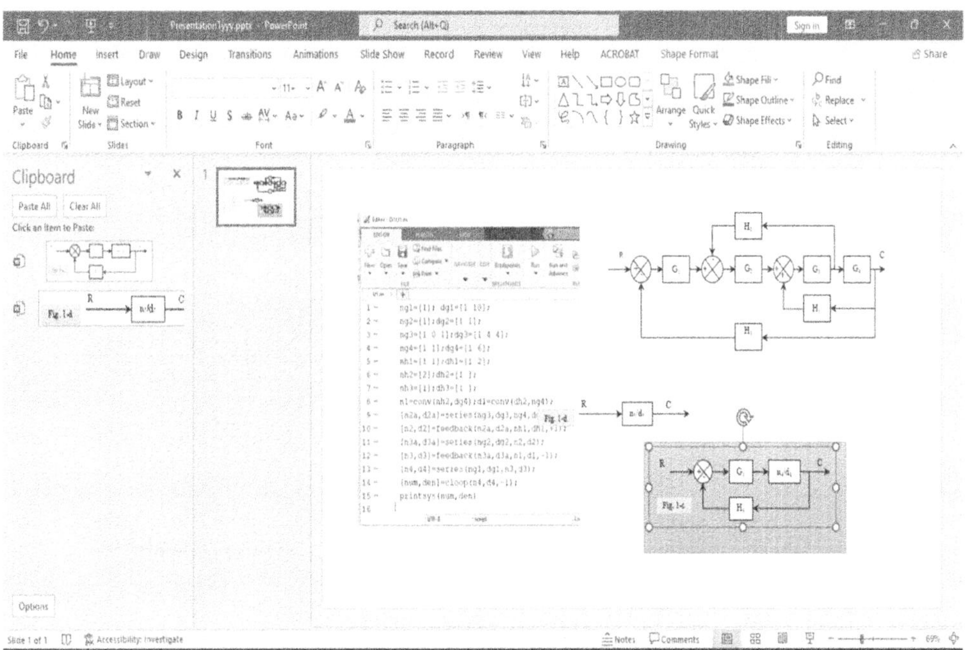

Figure 5.11 Importing Figs. 3.1-d and 3.1-c to PowerPoint slide.

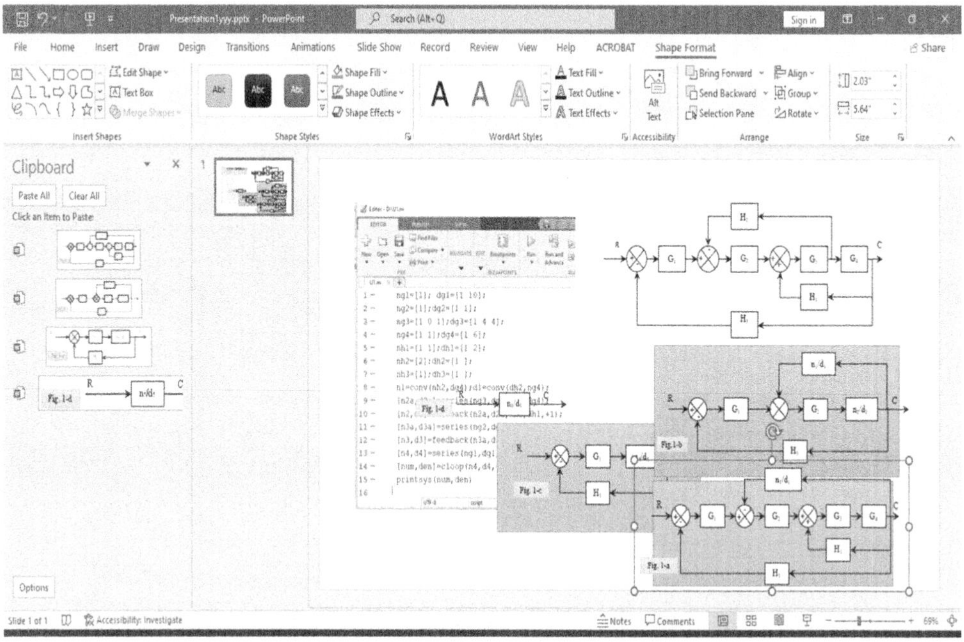

Figure 5.12 Importing Figs. 3.1-b and 3.1-a.

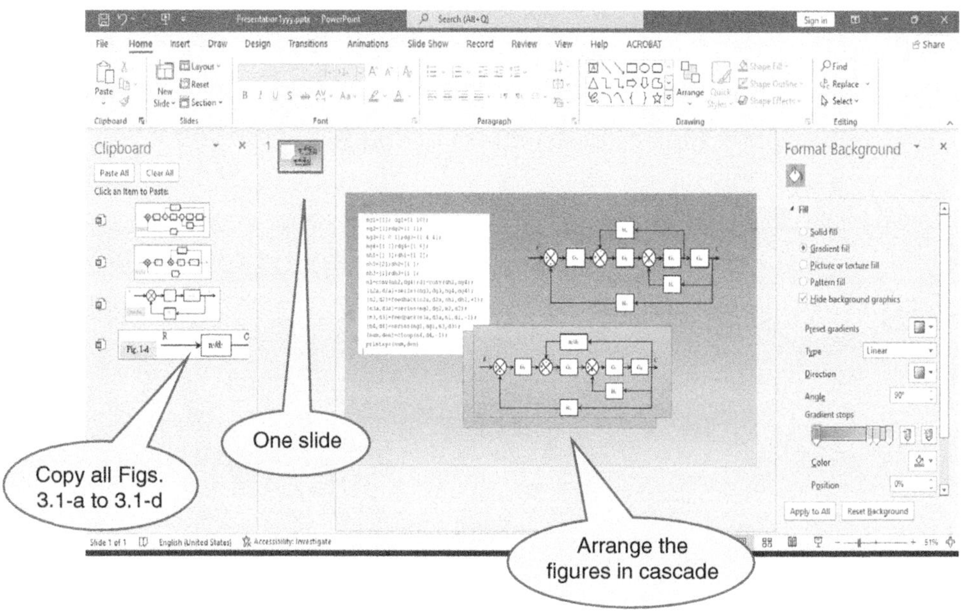

Figure 5.13 PowerPoint screen.

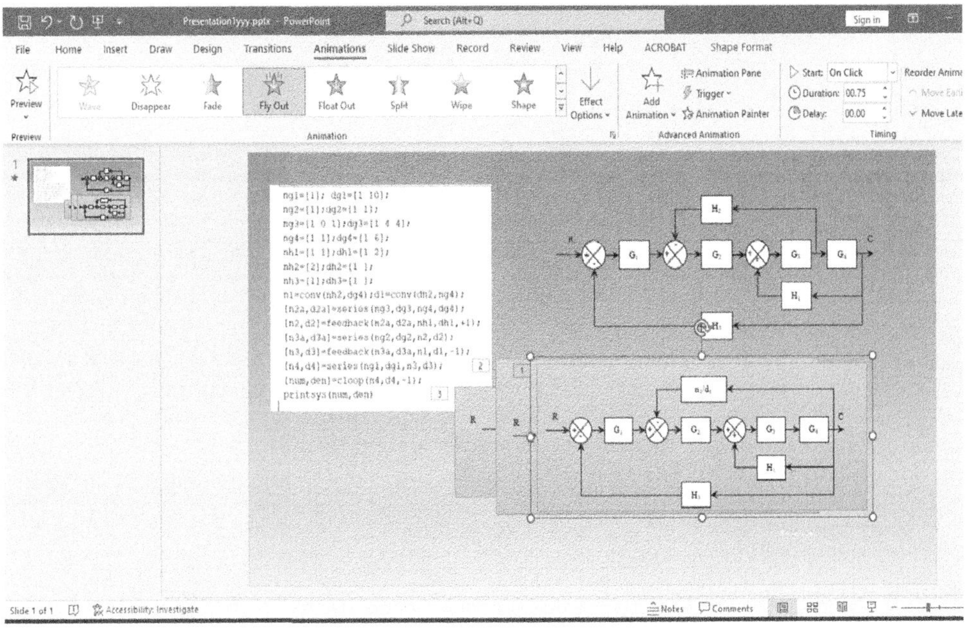

Figure 5.14 Animate specific figures.

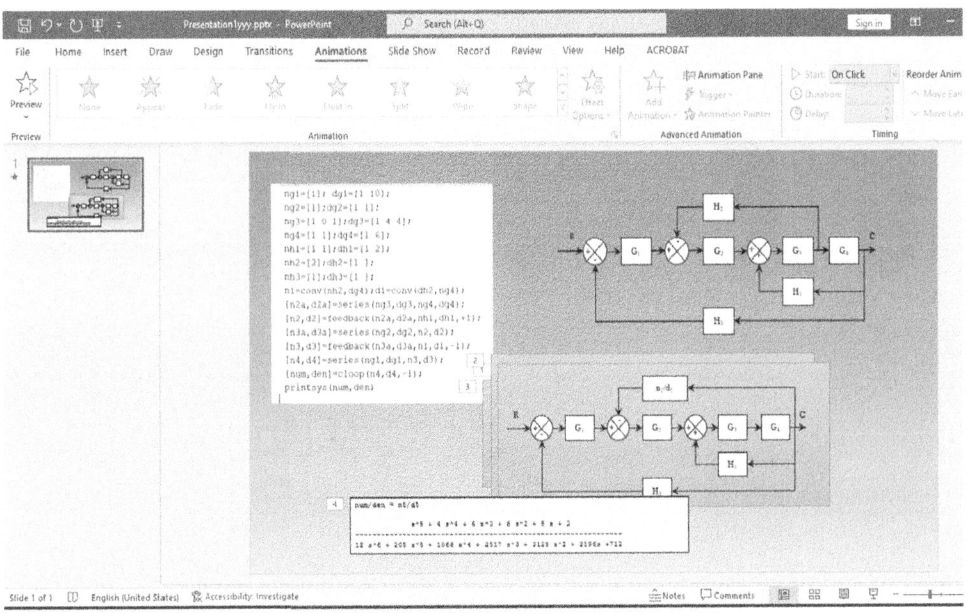

Figure 5.15 Display the final transfer function.

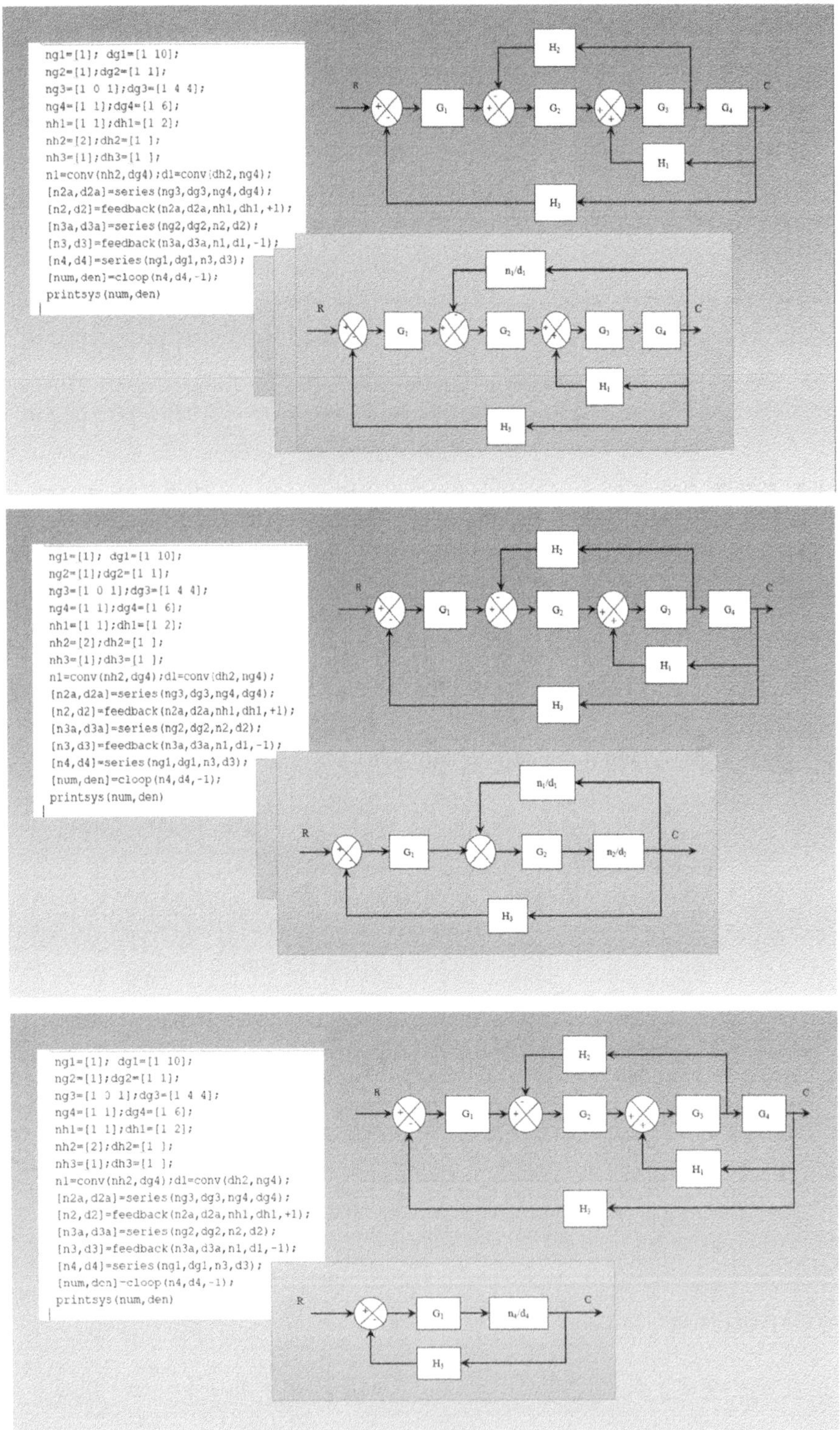

Figure 5.16 Presentation slide show.

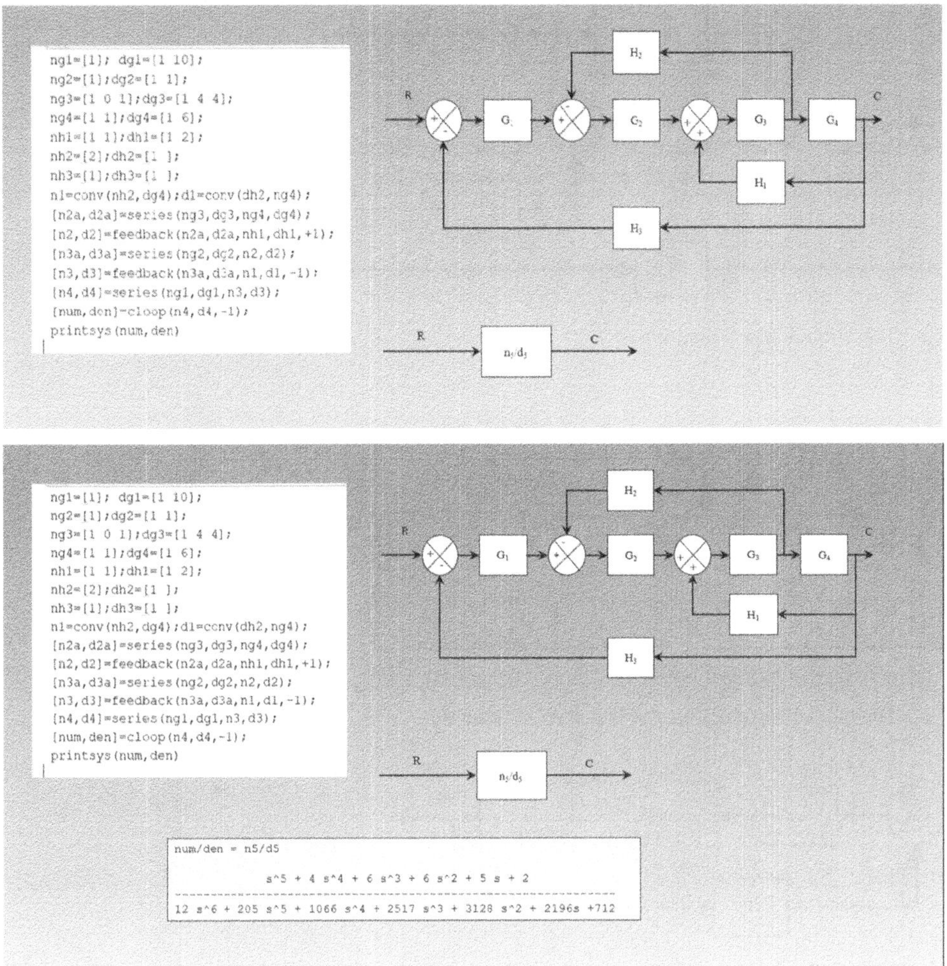

Figure 5.16 *(Continued)*

Crucial Note

Figures 5.17-a to 5.17-d should not be used in the presentation since they contain transfer functions in terms of G's and H's such as H_2/G_4, $G_3G_4/(1-G_3G_4H_1)$, and $G_2G_3G_4/(1-G_3G_4H_1+G_2G_3H_2)$. In contrast, the MATLAB program includes symbols such as n_1, d_1, n_2, and d_2. In this scenario, the software does not match the transfer function expression on the block diagram; thus, your audience will be confused and interrupt you during the slide show. Although Figs. 5.17-a through 5.17-d are all valid and correct, they do not match with the program.

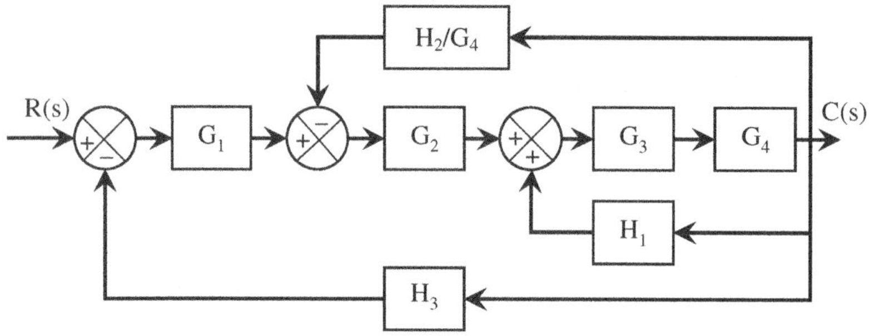

Figure 5.17-a The original block diagram.

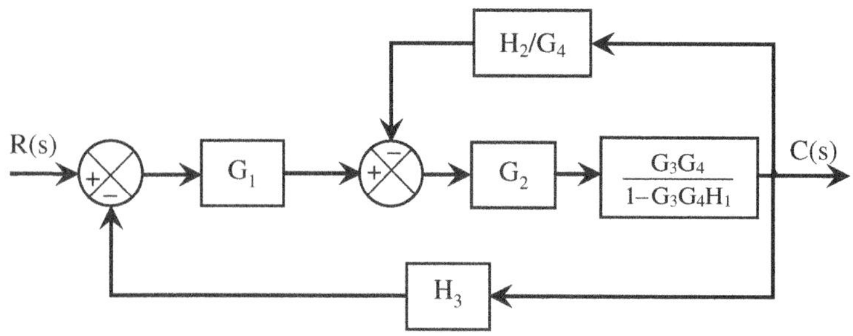

Figure 5.17-b The block diagram after the first step of reduction.

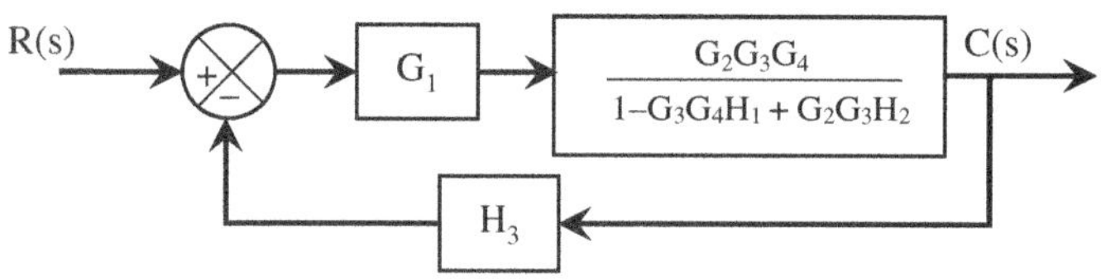

Figure 5.17-c The block diagram after the second step of reduction.

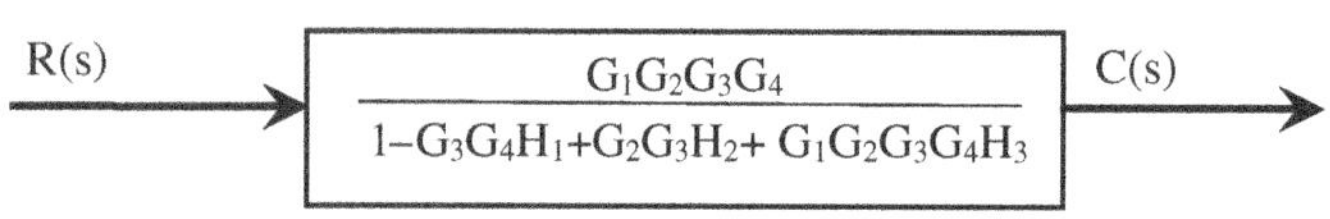

Figure 5.17-d The final block diagram.

Example 5.2 Another presentation idea for Example 5.1.

Process:

If you are not familiar with animations, follow the second method that was described previously.

1- Organize your slide as shown in Fig. 5.1.

2- On the first slide:
   - Place the MATLAB code in part 1.
   - Place the original block diagram (Fig. 3.1/Chapter 3) in part 2.
   - Place Fig. 3.1-a in part 3
   - Place the transfer function for this case in part 4 if you want to show it on the screen.
   - Your slide should look like Fig. 5.18
3- On the second slide, do the same as in the first slide except:
   - Place Fig. 3.1-b in part 3.
   - Place the transfer function for this case in part 4.
4- On the third slide, do the same as in the first slide except:
   - Place Fig. 3.1-c in part 3.
   - Place the transfer function for this case in part 4.
5- On the fourth slide, do the same as in the first slide except:
   - Place Fig. 3.1-d in part 3.
   - Place the final transfer function in part 4.
6- Set the background for each slide to that of the first slide.
7- No need for any animations.
8- Save your file.
9- On the slide show, after the completion of the explanation on the first slide, click the mouse or use the keyboard arrows to switch to the next slide. Do the same until finish as shown in Fig. 5.19.

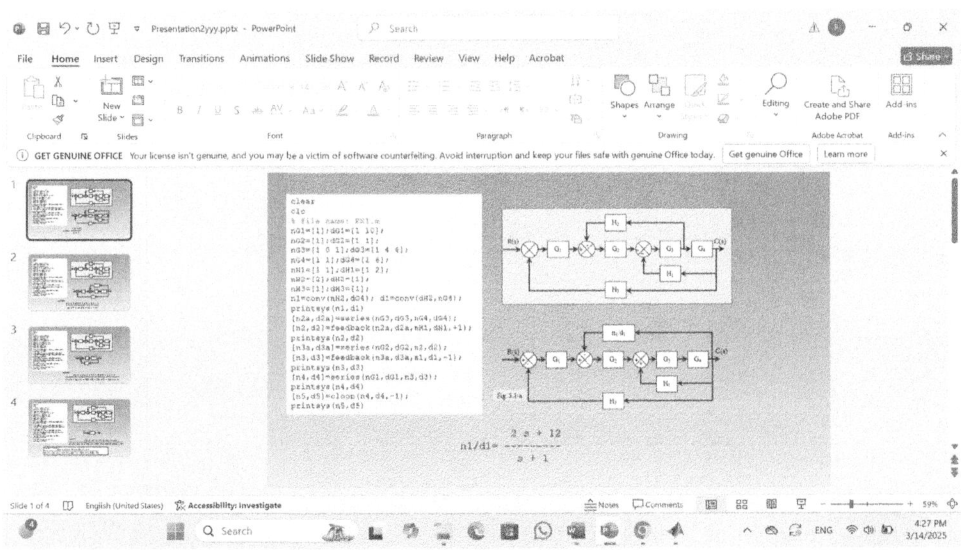

Figure 5.18 The first slide arrangement.

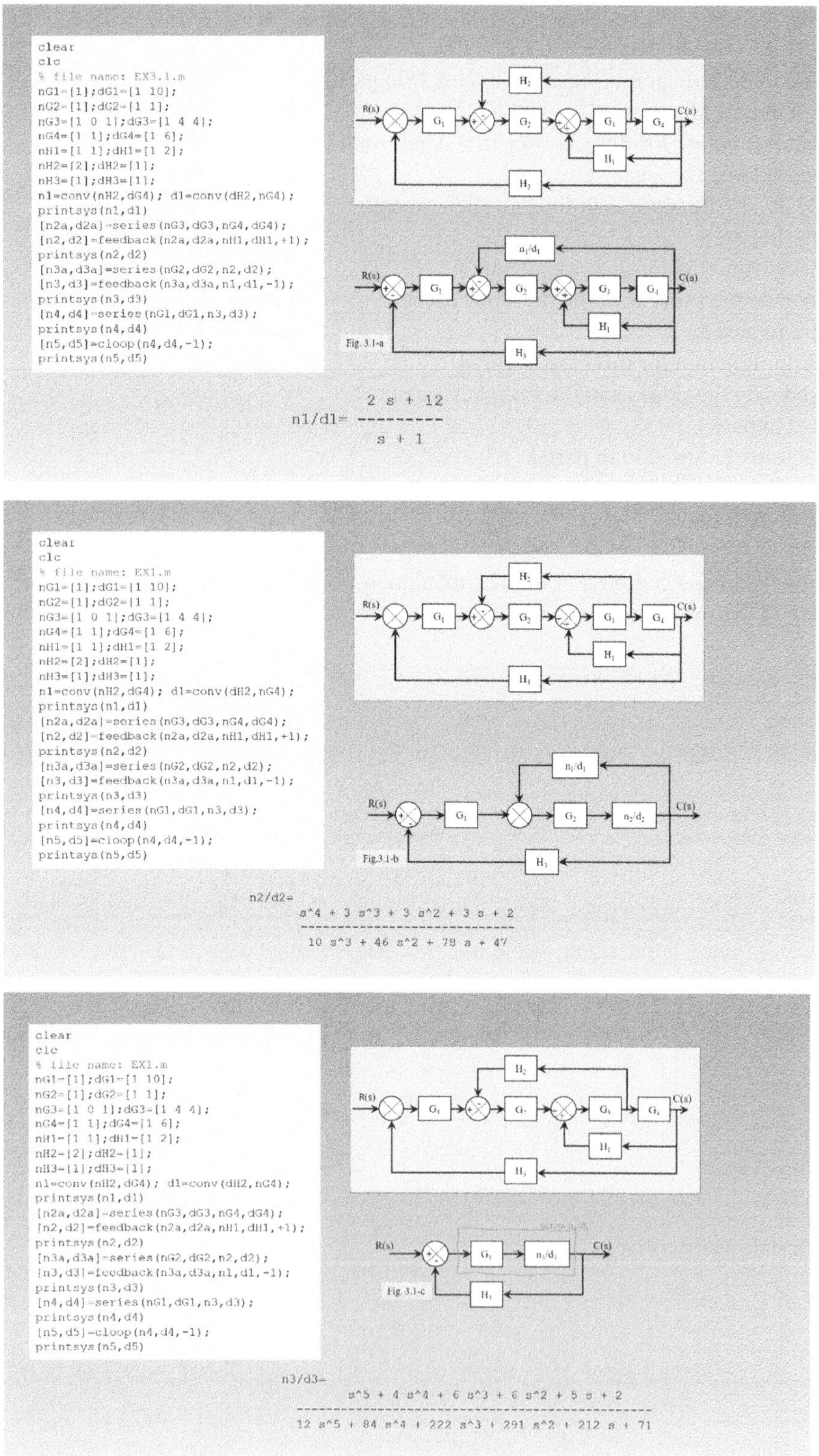

Figure 5.19 Presentation slide show of Example 5.2.

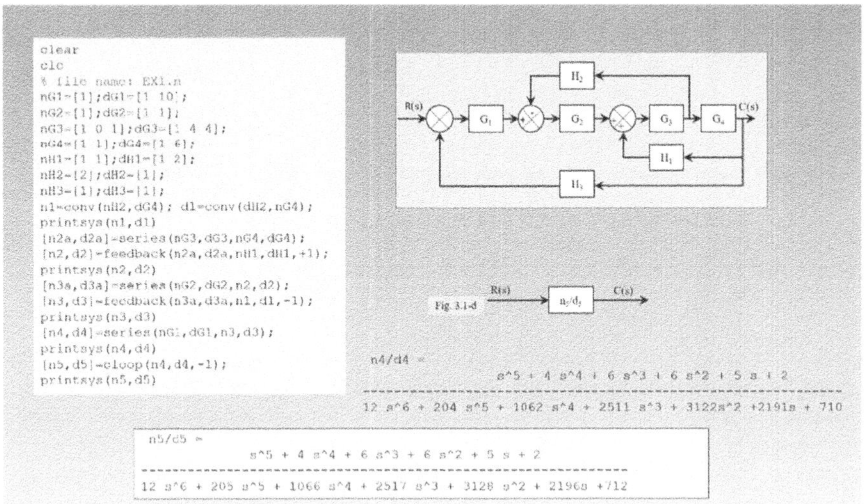

Figure 5.19 *(Continued)*

Activity:

Apply this method of presentation to another example.

Example 5.3 Present the block diagram of Fig. 3.4/Chapter 3 and show its reduction steps.

Process:

Because the block diagram of Fig. 3.4/Chapter 3 is simple, neat with repeatable reduction steps, we will follow the third method of presentation.

1- Open a New blank PowerPoint presentation file.
2- Go to Microsoft Word file that contains the block diagram of Fig. 3.4 and copy it.
3- Return to your PowerPoint file and paste the figure as shown in Fig. 5.20.
4- Mark the first feedback with a labeled circle and then animate it. Use, for example, the Zoom option from the Entrance category. Set the Duration to 1.75 as shown in Fig. 5.21.
5- Mark the first series with a labeled arc and animate it. Use, for example, the Fly In option from the Entrance category. Set the duration to 1 as shown in Fig. 5.21.
6- Repeat the same process for the rest feedback loops and series groups. Try to use the same animations with the same duration to keep your audience on track. In other words, to keep them connected.
7- Return to Microsoft Word file and copy the final transfer function, then paste it into your PowerPoint file. Animate it by using the Appear option from the Entrance category. Your presentation file should look like Fig. 5.22.

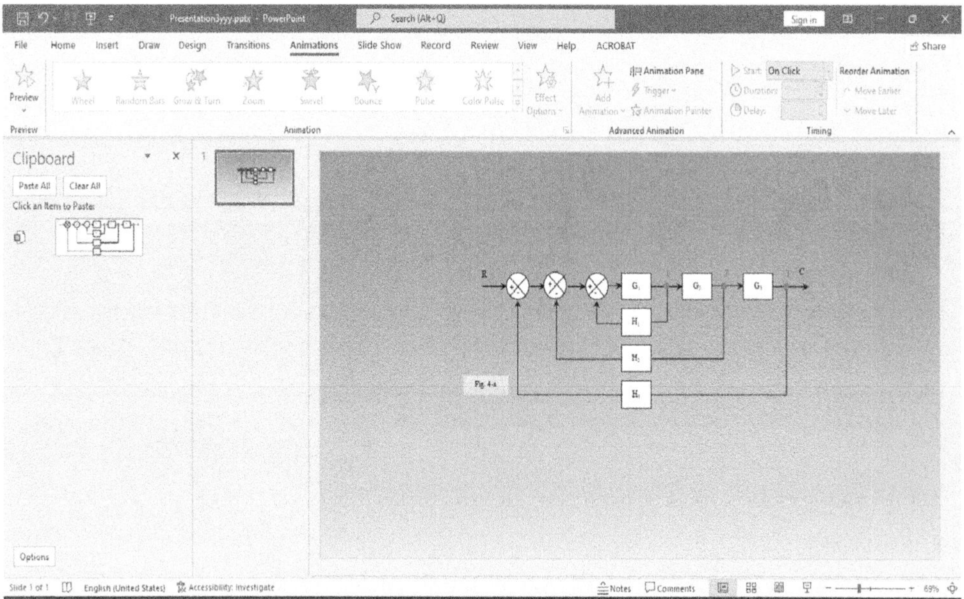

Figure 5.20 Importing Fig. 3.4/Chapter 3 into the presentation file.

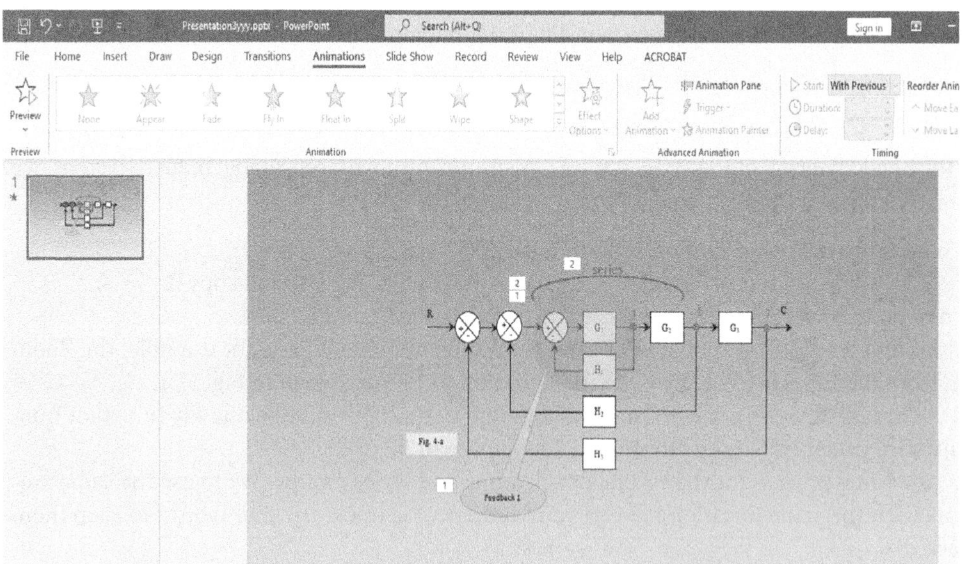

Figure 5.21 The presentation file after complete marking the steps 4 and 5.

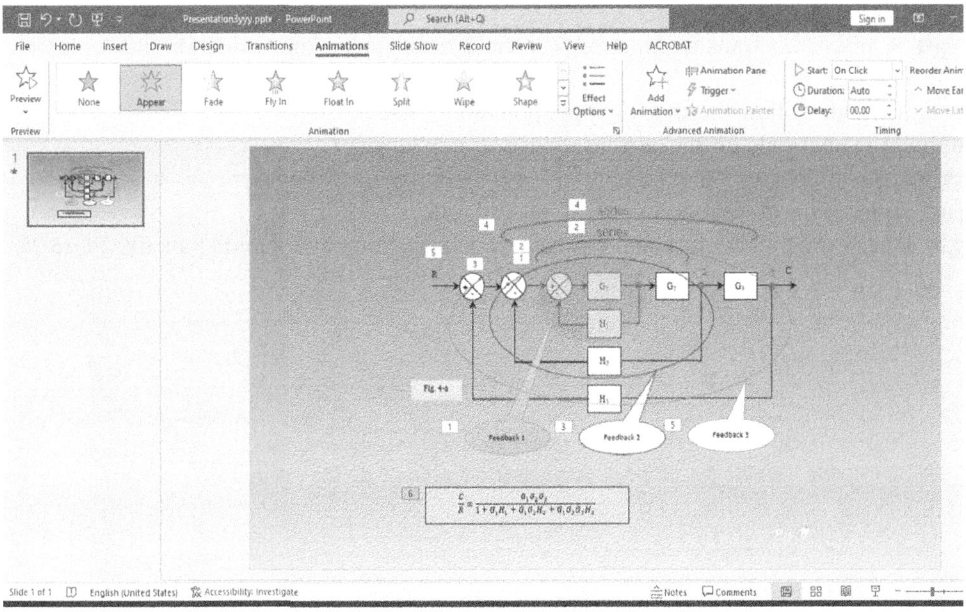

Figure 5.22 The final look of the presentation file.

Note:

We did not use many block diagrams in this section; instead, we just explained the reduction process. You can pick your presentation process based on your aim or focus.

Example 5.4 Present the block diagram of Fig. 3.6/Chapter 3, and show its reduction steps.

First Process (Without Animations):

By following the fourth method of presentation, the presentation file for the reduction steps of Fig. 3.6/Chapter 3 is structured as follows:

You will need four slides.

1- On the first slide, follow the slide organization of Fig. 5.2:
   - Place the starting statements of the MATLAB code in part 1.
   - Place the original block diagram (Fig. 3.6/Chapter 3) in part 2.
   - In part 3, place the redrawn block diagram (Fig. 3.6-a/Chapter 3).
   - Leave part 4 empty because the reduction steps haven't been started yet. Your slide should look like Fig. 5.23.

2- Moving to the second slide, follow the slide organization of Fig. 5.3:
   - Add the MATLAB code for the current reduction step in part 1.
   - Place the redrawn block diagram (Fig. 3.6-a/Chapter 3) in part 3.
   - In part 5, place Fig. 3.6-b/Chapter 3.
   - In part 4, place the transfer function for the current step. Your slide should look like Fig. 5.24.

3- On the third slide, follow the slide organization of Fig. 5.3:
   - Add the MATLAB code for the current reduction step in part 1.
   - Place Fig. 3.6-b/Chapter in part 3.

- In part 5, place Fig. 3.6-d/Chapter 3.
- In part 4, place the transfer function for the current step. Your slide should look like Fig. 5.25.

4- On the fourth slide, follow the slide organization of Fig. 5.3:
- Add the MATLAB code for the current reduction step in part 1.
- Place Fig. 3.6-d/Chapter 3 in part 3.
- In part 5, place Fig. 3.6-e/Chapter 3.
- In part 4, place the transfer function for the current step. Your slide should look like Fig. 5.26.

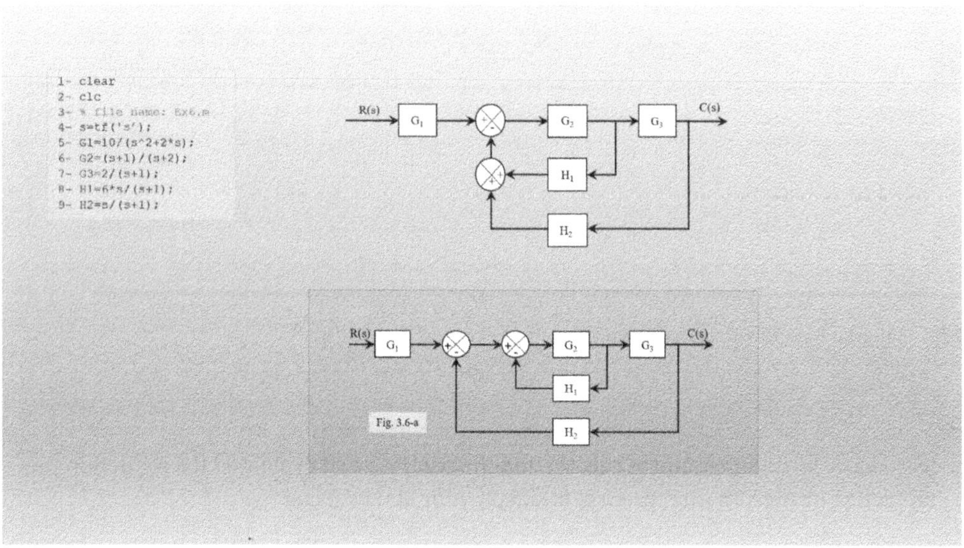

Figure 5.23 First slide arrangement of Example 5.4.

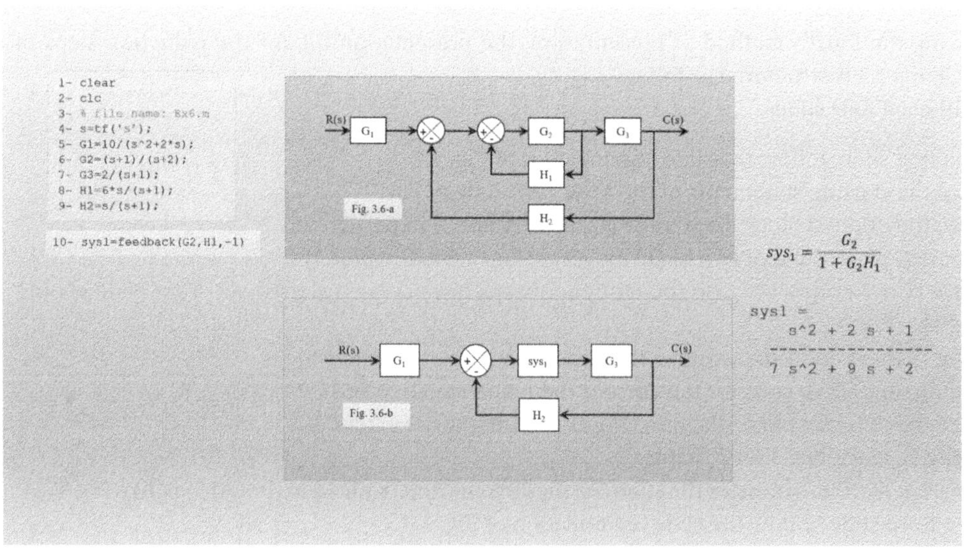

Figure 5.24 Second slide arrangement of Example 5.4.

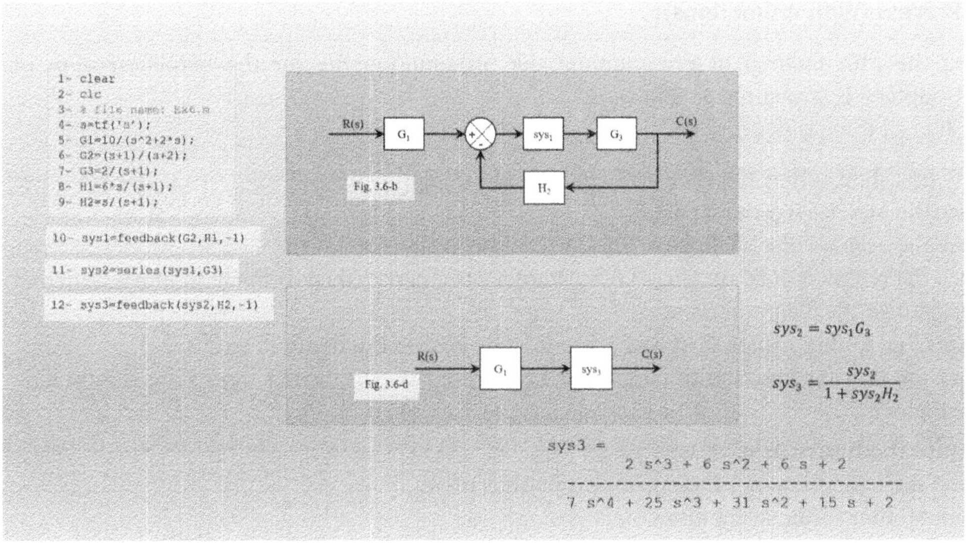

Figure 5.25 Third slide arrangement of Example 5.4.

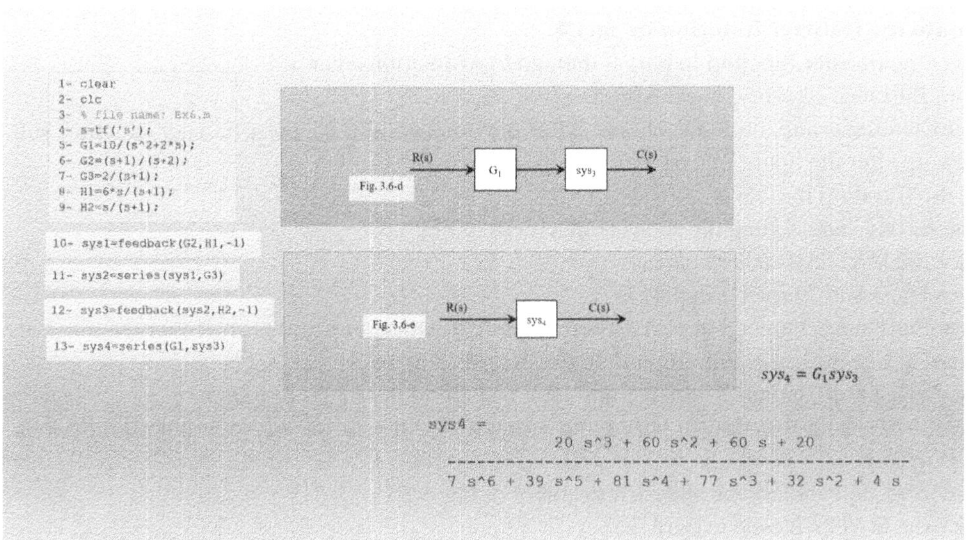

Figure 5.26 Fourth slide arrangement of Example 5.4.

Second Process (With Animations):

Following the fifth method of presentation, the presentation file for the reduction steps of Fig. 3.6/Chapter 3 is structured as follows:

You will need three slides in total.

1- On the first slide, follow the slide arrangement of Fig. 5.4:
- Place the MATLAB code in part 1.
- Place the original block diagram (Fig. 3.6/Chapter 3) in part 2.
- Place the redrawn diagram (Fig. 3.6-a/Chapter 3) in part 3, then animate it. Move it up and place it over the diagram in part 2.
- Place (Fig. 3.6-b/Chapter 3) in part 5 and let it be behind the figure in part 3.
- Place the transfer function in part 4 and animate it. After finishing, your slide should look like Fig. 5.27.

Animate the figure in part 3,
- Select the figure in part 3 then go to Animations menu.
- From Motion Paths, select Lines.
- On Effect Options, select Up. You will have a line with start and end points, drag and drop these points to adjust the path and the position of your figure.
- Go to Timing group, on Start, choose on click option. So, your figure will show up after a mouse click.
- Set the Duration to 2.

Animate the transfer function in part 4,
- Select the transfer function in part 4, then go to Animations menu.
- From Entrance category, select Appear.
- Go to Timing group, on Start, choose After previous option. So, your transfer function will show up after the figure in part 3 moves up.
- Set the Duration to 1.25.

2- On the second slide, follow the slide arrangement of Fig. 5.5:
- Place the MATLAB code in part 1.
- Place Fig. 3.6-a/Chapter 3 in part 2.
- Place Fig. 3.6-b/Chapter 3 in part 3, then animate it. Move it up and place it over the diagram of part 2. Use the same animation that was described above.
- Place Fig. 3.6-d/Chapter 3 in part 5 and let it be behind the figure in part 3.
- Place the transfer function in part 4 and animate it. After finishing, your slide should look like Fig. 5.28.

3- On the third slide, follow the slide arrangement of Fig. 5.5:
- Place the MATLAB code in part 1.
- Place Fig. 3.6-b/Chapter 3 in part 2.
- Place Fig. 3.6-d/Chapter 3 in part 3, then animate it. Move it up and place it over the diagram of part 2. Use the same animation that was described previously.
- Place Fig. 3.6-e/Chapter 3 in part 5 and let it be behind the figure in part 3.
- Place the transfer function in part 4 and animate it. After finishing, your slide should look like Fig. 5.29.

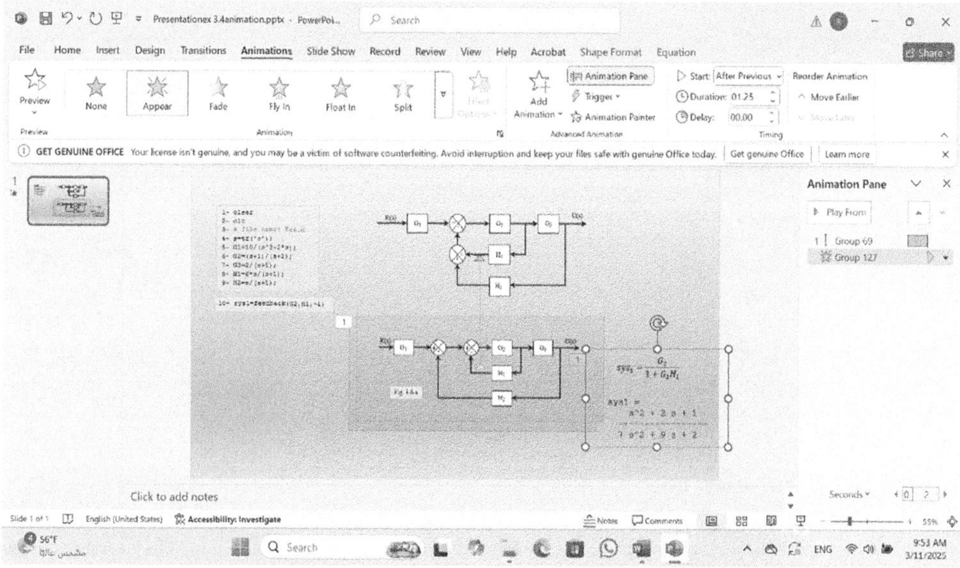

Figure 5.27 First slide arrangement.

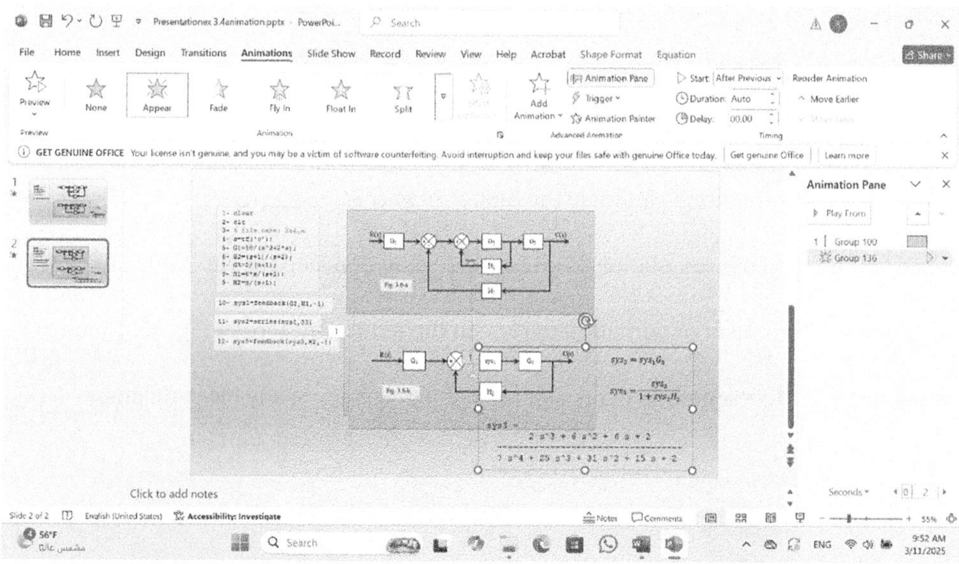

Figure 5.28 Second slide arrangement.

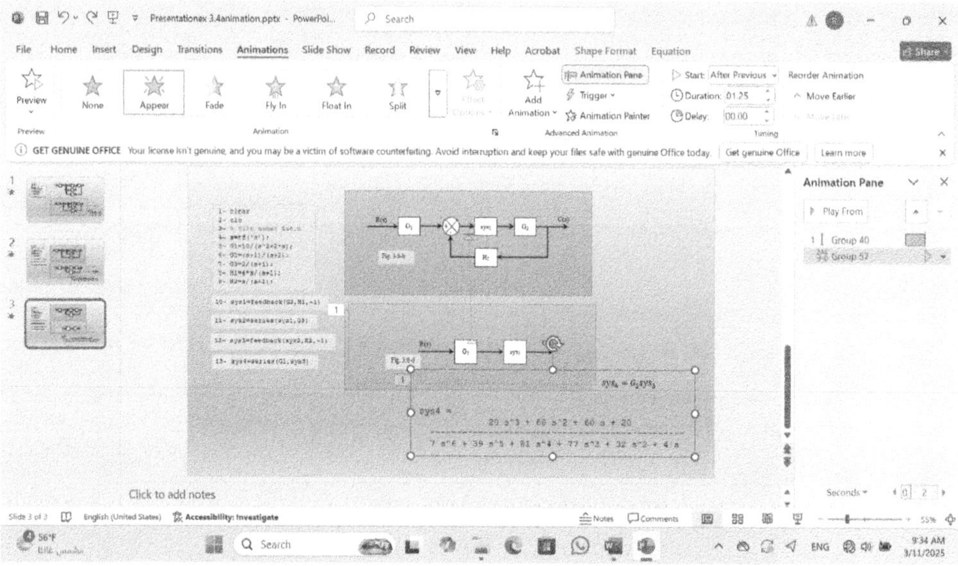

Figure 5.29 The last slide arrangement.

Review Questions

1 How can you import a MATLAB code into a presentation file?

2 Create another method for organizing a slide.

3 Use another type of animation to animate the figures in part 3.

4 Follow the first method of presentation to present the reduction steps of Fig. 3.2/Chapter 3.

5 Follow the second method of presentation to present the reduction steps of Fig. 3.5/Chapter 3.

6 Follow any method of presentation to present the reduction steps of any block diagram.

6

Signal Flow Graph

6.1 Introduction

Signal flow graph (SFG) is an alternative approach to solve complex block diagram problems that are hardly being solved by using the block diagram reduction rules. Signal flow graph is also called a Mason graph after Samuel J. Mason, who coined its ideology; SFG is a node-branch representation of simultaneous equations that describes a system. System variables are represented by Nodes in SFG, while the signal flow between them is represented by branches. By following simple rules, the overall transfer function can be obtained.

This chapter defines SFG and its terminology. Then Mason's theory is explained and discussed. Different examples are taken to illustrate how the overall transfer function can be obtained. The last section demonstrates how to convert a block diagram into SFG.

Learning Objectives:

- To learn the aspect of the SFG.
- To learn how to solve the SFG.
- To learn how to apply Mason's theorem to the SFG.
- To learn how to convert a block diagram into SFG.

6.2 Signal Flow Graph (SFG)

SFG is a pictorial representation of a control system, and it is formed by nodes and branches. Each node represents a variable, and each branch represents the line that connects two nodes. Figure 6.1 shows SFG for a control system [2, p: 106].

Block Diagram Implementation for Control Systems: Including MATLAB® Programming, Microsoft® Word and PowerPoint Presentation, First Edition. Ruba Al-Mulla Hummadi.

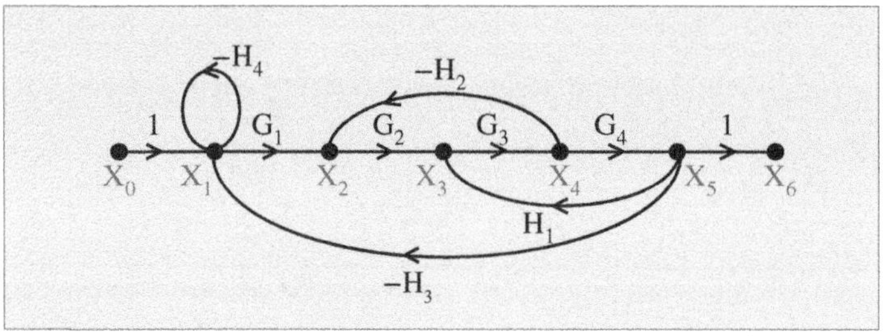

Figure 6.1 Signal flow graph.

6.3 Terminology in SFG

Table 6.1 describes all the nodes, paths, and loop types of SFG of Fig. 6.1. On the other hand, Fig. 6.2 shows the terminology of SFG of Fig. 6.1.

Table 6.1 Node, path, and loop types in SFG.

Node type	**Description**	**Example**
Source node	Has only outgoing branches	X_0
Sink node	Has only incoming branches	X_6
Chain node	Has both incoming and outgoing branches	X_1, X_2, X_3, X_4, and X_5
Path type	**Description**	**Example**
Forward path	A path from the input to the output without tracing the same node twice.	X_0 - X_1 - X_2 - X_3 - X_4 - X_5 - X_6
Loop type	**Description**	**Example**
Feedback loop	A loop that starts and ends at the same node.	X_1 - X_2 - X_3 - X_4 - X_5 - X_1 1st loop X_2 - X_3 - X_4 - X_2 2nd loop X_3 - X_4 - X_5 - X_3 3rd loop
Self-loop	Feedback loop that originates and terminates at the same node without passing another node.	X_1 4th loop
Non-touching loop	Loops that don't share any common node.	2nd loop with 4th loop 3rd loop with 4th loop

6.4 SFG Algebra

Any complex signal flow graph can be reduced by using SFG reduction algebra rules. These rules are all tabulated in Jairth [2, p: 108, Table 5.2] and categorized into different operations such as addition, multiplication, elimination of feedback loop, and miscellaneous. Because this is a tedious process and takes a long time, Mason was the first who coined the term for his gain formula. Therefore, the transfer function can be acquired easily.

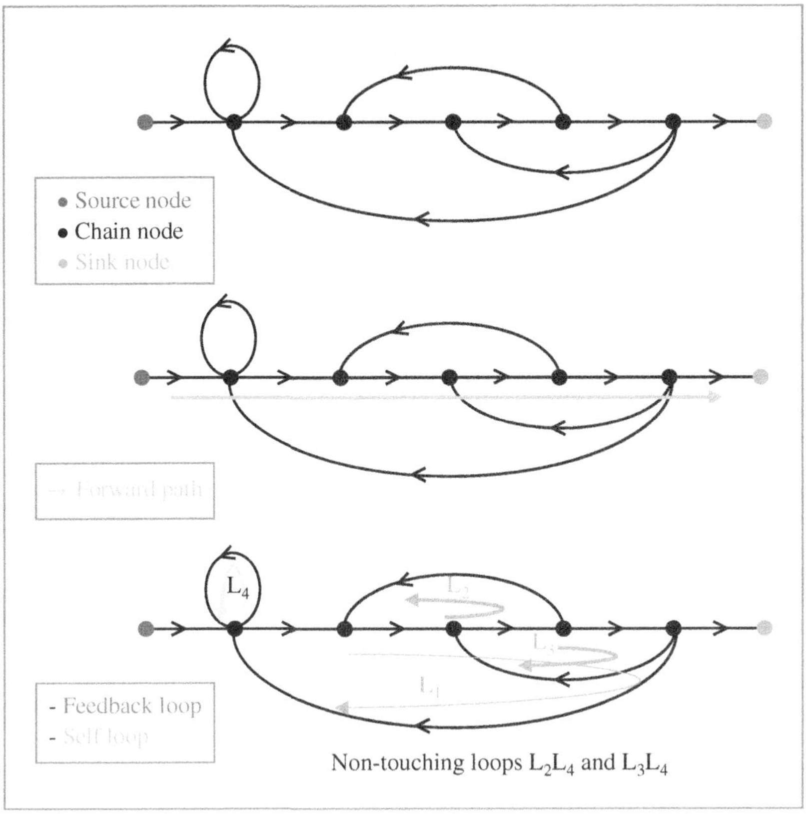

Figure 6.2 SFG terminology illustration.

6.5 Mason's Gain Formula [2, p: 108–109]

By applying the following formula, the transfer function of a signal flow graph can be obtained:

$$T = \frac{\sum_{K=1}^{n} P_K \Delta_K}{\Delta}$$

where

T is the overall system transfer function.
K is the forward paths counter.
n is the maximum number of forward paths.
P_K is the Kth forward path gain.
Δ is the determinant of the signal flow graph. Or,
$\Delta = 1 -$ (sum of all individual loops gains) + (sum of the gain products of all possible two non-touching loops) − (sum of the gain products of all possible three non-touching loops) + (…) − (…) +
Δ_K is the cofactor of the kth forward path. Or,
Δ_K is the same as Δ but excluding the loops that are attached to the kth forward path. Or,

$\Delta_K = 1 -$ (sum of gain of all individual loops that are isolated from the kth forward path) + (sum of the gain products of all possible two non-touching loops that are isolated from the kth forward path) − (sum of the gain products of all possible three non-touching loops that are isolated from the kth forward path) + (....) − (....) +

6.6 Examples of SFG

Example 6.1 Obtain C(s)/R(s) for the signal flow graph of Fig. 6.3.

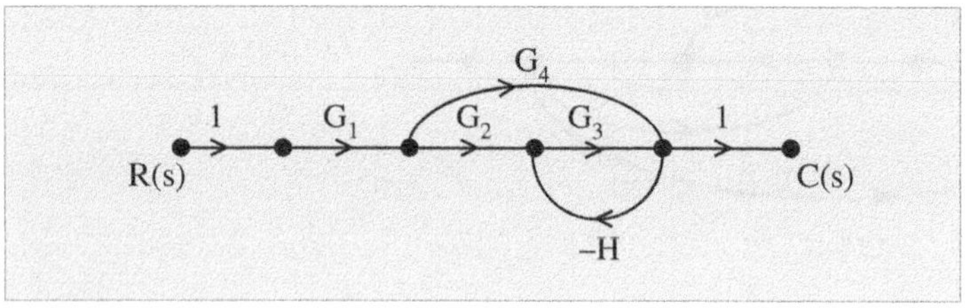

Figure 6.3 The signal flow graph of Example 6.1.

Solution:

The forward paths' gains:

$P_1 = G_1G_2G_3$ k: number of forward paths = 2

$P_2 = G_1G_4$

Individual loops:

$L = -G_3H$

Two non-touching loops:

Not exist in this example.

Three non-touching loops:

Not exist in this example.

$$\frac{C(s)}{R(s)} = \frac{\sum_{k=1}^{2} P_k\Delta_k}{\Delta} = \frac{P_1\Delta_1 + P_2\Delta_2}{\Delta}$$

$$\Delta_1 = 1 - 0 = 1$$

$$\Delta_2 = 1 - 0 = 1$$

$$\Delta = 1 - L = 1 + G_3H$$

$$P_1\Delta_1 = G_1G_2G_3, \quad P_2\Delta_2 = G_1G_4$$

$$\frac{C(s)}{R(s)} = \frac{G_1G_2G_3 + G_1G_4}{1 + G_3H}$$

Activity:

Obtain $Y_1(s)/U(s)$ for the system of Fig. 6.4 [2, p: 206, Fig. 7.33].

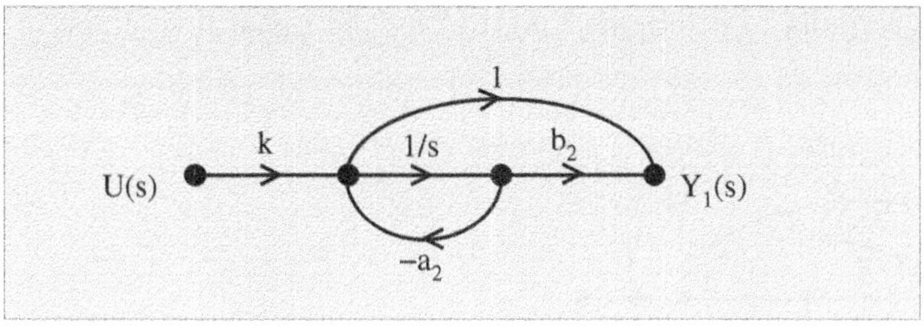

Figure 6.4 The signal flow graph of Example 6.1 activity. Jairth [2] / Ane Books.

Example 6.2 For the signal flow graph of Fig. 6.1, find the overall system transfer function by using Mason's gain formula.

Solution:

The forward path gain:

$P_1 = G_1G_2G_3G_4$ k: number of forward paths = 1

Individual loops:

$$L_1 = -G_1G_2G_3G_4H_3$$
$$L_2 = -G_2G_3H_2$$
$$L_3 = G_3G_4H_1$$
$$L_4 = -H_4$$

Two non-touching loops:

$$L_2L_4 = G_2G_3H_2H_4$$
$$L_3L_4 = -G_3G_4H_1H_4$$

Three non-touching loops:

Not exist in this example.

$$\Delta_1 = 1 - 0 = 1$$
$$\Delta = 1 - (L_1 + L_2 + L_3 + L_4) + (L_2L_4 + L_3L_4) - 0$$
$$\Delta = 1 - (-G_1G_2G_3G_4 + (-G_2G_3H_2) + G_3G_4H_1 - H_4) + \left(G_2G_3H_2H_4\right) + (-G_3G_4H_1H_4)$$
$$P_1\Delta_1 = G_1G_2G_3G_4$$
$$T(s) = \frac{P_1\Delta_1}{\Delta}$$

$$\boxed{T(s) = \frac{G_1G_2G_3G_4}{1 + G_1G_2G_3G_4 + G_2G_3H_2 - G_3G_4H_1 + H_4 + G_2G_3H_2H_4 - G_3G_4H_1H_4}}$$

For More Practice:

See Example 7 [1, p: 18].

Example 6.3 Use Mason's gain formula to determine C(s)/R(s) for the signal flow graph of Fig. 6.5.

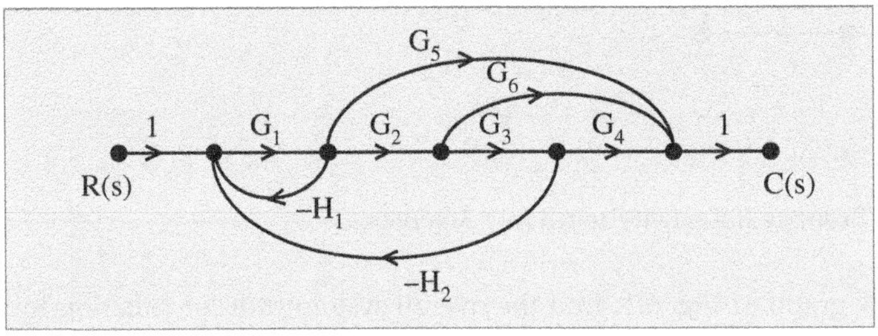

Figure 6.5 The signal flow graph of Example 6.3.

Solution:

The forward paths' gains:

$P_1 = G_1G_2G_3G_4$ k: number of forward paths = 3

$P_2 = G_1G_5$

$P_3 = G_1G_2G_6$

Individual loops:

$L_1 = -G_1H_1$

$L_2 = -G_1G_2G_3H_2$

Two non-touching loops

Not exist in this example.

Three non-touching loops

Not exist in this example.

$$\frac{C(s)}{R(s)} = \frac{\sum_{k=1}^{3} P_k\Delta_k}{\Delta} = \frac{P_1\Delta_1 + P_2\Delta_2 + P_3\Delta_3}{\Delta}$$

$$\Delta_1 = \Delta_2 = \Delta_3 = 1 - 0 = 1$$

$$\Delta = 1 - (L_1 + L_2) = 1 + G_1H_1 + G_1G_2G_3H_2$$

$$P_1\Delta_1 = G_1G_2G_3G_4, \quad P_2\Delta_2 = G_1G_5, \quad P_3\Delta_3 = G_1G_2G_6$$

$$\boxed{\frac{C(s)}{R(s)} = \frac{G_1G_2G_3G_4 + G_1G_5 + G_1G_2G_6}{1 + G_1H_1 + G_1G_2G_3H_2}}$$

For More Practice:

See Problems 5.2 and 5.3 [2, p: 111–113] and ex. 8, 9, and 10 [1, p: 19–21].

Example 6.4 Obtain C(s)/R(s) for the signal flow graph of Fig. 6.6 [2, p: 121, problem 16/b].

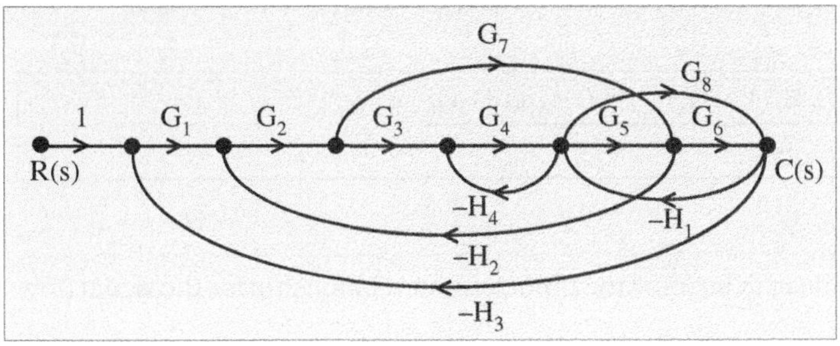

Figure 6.6 The signal flow graph of Example 6.4. Jairth [2] / Ane Books.

Solution:

The forward paths' gains:

$P_1 = G_1G_2G_3G_4G_5G_6$ k: number of forward paths = 3

$P_2 = G_1G_2G_7G_6$

$P_3 = G_1G_2G_3G_4G_8$

Individual loops:

$L_1 = -G_4H_4$

$L_2 = -G_5G_6H_1$

$L_3 = -G_2G_3G_4G_5H_2$

$L_4 = -G_1G_2G_3G_4G_5G_6H_3$

$L_5 = -G_8H_1$

$L_6 = -G_2G_7H_2$

$L_7 = -G_1G_2G_7G_6H_3$

$L_8 = -G_1G_2G_3G_4G_8H_3$

Two non-touching loops:

$L_1L_6 = G_4H_4G_2G_7H_2$

$L_1L_7 = G_4H_4G_1G_2G_7G_6H_3$

$L_5L_6 = G_8H_1G_2G_7H_2$

Three non-touching loops:

Not exist in this example.

$$\frac{C(s)}{R(s)} = \frac{\sum_{k=1}^{3} P_k\Delta_k}{\Delta} = \frac{P_1\Delta_1 + P_2\Delta_2 + P_3\Delta_3}{\Delta}$$

$\Delta_1 = 1$

$$\Delta_2 = 1 - L_1 = 1 + G_4H_4$$

$$\Delta_3 = 1$$

$$\Delta = 1 - (L_1 + L_2 + L_3 + L_4 + L_5 + L_6 + L_7 + L_8) + (L_1L_6 + L_1L_7 + L_5L_6)$$

$$P_1\Delta_1 = G_1G_2G_3G_4G_5G_6, \quad P_2\Delta_2 = G_1G_2G_7G_6(1 + G_4H_4), \quad P_3\Delta_3 = G_1G_2G_3G_4G_8$$

$$\frac{C(s)}{R(s)} = \frac{G_1G_2G_3G_4G_5G_6 + G_1G_2G_7G_6(1 + G_4H_4) + G_1G_2G_3G_4G_8}{\Delta}$$

Example 6.5 Use Mason's formula to determine the input/output relationship for the signal flow graph of Fig. 6.7.

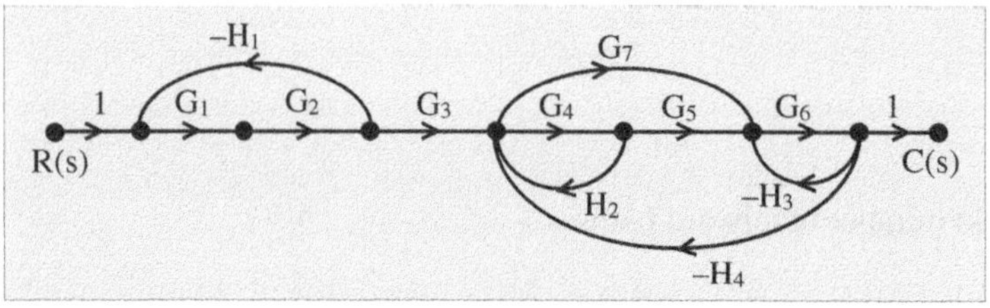

Figure 6.7 The signal flow graph of Example 6.5.

Solution:

The forward paths' gains:

$P_1 = G_1G_2G_3G_4G_5G_6$ k: number of forward paths = 2

$P_2 = G_1G_2G_3G_7G_6$

Individual loops:

$L_1 = -G_1G_2H_1$

$L_2 = G_4H_2$

$L_3 = -G_6H_3$

$L_4 = -G_4G_5G_6H_4$

$L_5 = -G_7G_6H_4$

Two non-touching loops:

$L_1L_2 = -G_1G_2G_4H_1H_2$

$L_1L_3 = G_1G_2G_6H_1H_3$

$L_1L_4 = G_1G_2G_4G_5G_6H_1H_4$

$L_1L_5 = G_1G_2G_7G_6H_1H_4$

$L_2L_3 = -G_4G_6H_2H_3$

Three non-touching loops:

$$L_1L_2L_3 = G_1G_2G_4G_6H_1H_2H_3$$

$$\frac{C(s)}{R(s)} = \frac{\sum_{k=1}^{2} P_k\Delta_k}{\Delta} = \frac{P_1\Delta_1 + P_2\Delta_2}{\Delta}$$

$$\Delta_1 = 1$$

$$\Delta_2 = 1$$

$$\Delta = 1 - (L_1 + L_2 + L_3 + L_4 + L_5) + (L_1L_2 + L_1L_3 + L_1L_4 + L_1L_5 + L_2L_3) - (L_1L_2L_3)$$

$$\begin{aligned}\Delta = 1 &- (-G_1G_2H_1 + G_4H_2 - G_6H_3 - G_4G_5G_6H_4 - G_7G_6H_4)\\ &+ \big(-G_1G_2G_4H_1H_2 + G_1G_2G_6H_1H_3 + G_1G_2G_4G_5G_6H_1H_4 + G_1G_2G_7G_6H_1H_4\\ &- G_4G_6H_2H_3\big) - G_1G_2G_4G_6H_1H_2H_3\end{aligned}$$

$$P_1\Delta_1 = G_1G_2G_3G_4G_5G_6, \quad P_2\Delta_2 = G_1G_2G_3G_7G_6$$

$$\boxed{\frac{C(s)}{R(s)} = \frac{G_1G_2G_3G_6(G_4G_5 + G_7)}{1 - G_4H_2 + G_6H_3 + (1 + G_9)(G_8 + G_7G_6H_4 - G_{10}) + G_9(1 - G_4H_2 + G_6H_3)}}$$

Where $G_8 = G_4G_5G_6H_4$, $G_9 = G_1G_2H_1$, $G_{10} = G_4G_6H_2H_3$

For More Practice:

Solve the signal flow graphs of Fig. 7.58-e and Fig. 8.13 [2, p: 236 & 300].

Example 6.6 Use Mason's formula to determine the input/output relationship for the signal flow graph of Fig. 6.8, knowing that the idea of this question originated from Jairth [2, p: 168, Fig. 7-10].

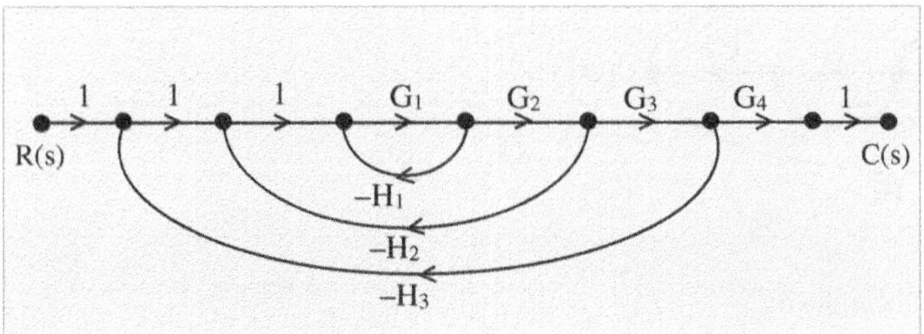

Figure 6.8 The signal flow graph of Example 6.6. Jairth [2] / Ane Books.

Solution:

The forward path gain:

$P_1 = G_1G_2G_3G_4$ k: number of forward paths = 1

Individual loops:

$$L_1 = -G_1H_1$$

$$L_2 = -G_1G_2H_2$$

$$L_3 = -G_1G_2G_3H_3$$

Two non-touching loops:

Not exist in this example.

Three non-touching loops:

Not exist in this example.

$$\Delta_1 = 1$$

$$\Delta = 1 - (L_1 + L_2 + L_3)$$

$$\Delta = 1 - (-G_1H_1 - G_1G_2H_2 - G_1G_2G_3H_3)$$

$$\Delta = 1 + G_1H_1 + G_1G_2H_2 + G_1G_2G_3H_3$$

$$P_1\Delta_1 = G_1G_2G_3G_4$$

$$\frac{C(s)}{R(s)} = \frac{P_1\Delta_1}{\Delta}$$

$$\frac{C(s)}{R(s)} = \frac{G_1G_2G_3G_4}{1 + G_1H_1 + G_1G_2H_2 + G_1G_2G_3H_3}$$

Example 6.7 Use Mason's formula to determine the C(s)/R(s) relationship for the signal flow graph of Fig. 6.9.

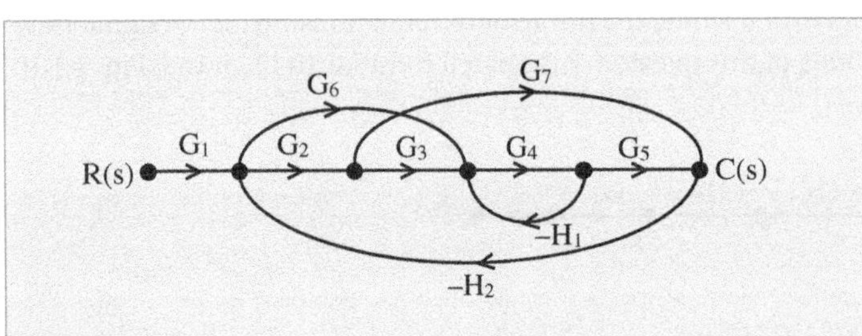

Figure 6.9 The signal flow graph of Example 6.7.

Solution:

The forward paths' gains:

$P_1 = G_1G_2G_3G_4G_5$ k: number of forward paths = 3

$P_2 = G_1G_6G_4G_5$

$P_3 = G_1G_2G_7$

Individual loops:

$L_1 = -G_4H_1$

$L_2 = -G_2G_3G_4G_5H_2$

$$L_3 = -G_6G_4G_5H_2$$

$$L_4 = -G_2G_7H_2$$

Two non-touching loops:

$$L_1L_4 = G_4G_2G_7H_1H_2$$

Three non-touching loops:

Not exist in this example.

$$\frac{C(s)}{R(s)} = \frac{\sum_{k=1}^{3} P_k\Delta_k}{\Delta} = \frac{P_1\Delta_1 + P_2\Delta_2 + P_3\Delta_3}{\Delta}$$

$$\Delta = 1 - (L_1 + L_2 + L_3 + L_4) + L_1L_4$$

$$\Delta = 1 - (-G_4H_1 - G_2G_3G_4G_5H_2 - G_6G_4G_5H_2 - G_2G_7H_2) + G_4G_2G_7H_1H_2$$

$$\Delta_1 = 1 - 0$$

$$\Delta_2 = 1$$

$$\Delta_3 = 1 - L_1 = 1 + G_4H_1$$

$$P_1\Delta_1 = G_1G_2G_3G_4G_5$$

$$P_2\Delta_2 = G_1G_6G_4G_5$$

$$P_3\Delta_3 = G_1G_2G_7(1 + G_4H_1)$$

$$\frac{C(s)}{R(s)} = \frac{G_1G_2G_3G_4G_5 + G_1G_6G_4G_5 + G_1G_2G_7(1 + G_4H_1)}{1 + G_4H_1 + G_2G_3G_4G_5H_2 + G_6G_4G_5H_2 + G_2G_7H_2 + G_4G_2G_7H_1H_2}$$

For More Practice:

See example 2.8 on page 66 [10].

Example 6.8 Obtain C(s)/R(s) for the signal flow graph of Fig. 6.10. Note that the block diagram of this signal flow graph exists in Ogata [3, p: 47].

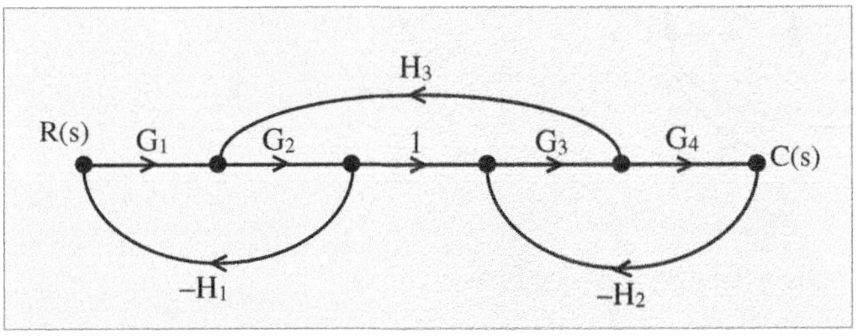

Figure 6.10 The signal flow graph of Example 6.8. Adapted from Ogata [3].

Solution:

The forward path gain:

$P_1 = G_1G_2G_3G_4$ k: number of forward paths = 1

Individual loops:

$L_1 = -G_1G_2H_1$

$L_2 = -G_3G_4H_2$

$L_3 = G_2G_3H_3$

Two non-touching loops:

$L_1L_2 = G_1G_2G_3G_4H_1H_2$

Three non-touching loops:

Not exist in this example.

$$\Delta = 1 - (L_1 + L_2 + L_3) + L_1L_2$$

$$\Delta = 1 - (-G_1G_2H_1 - G_3G_4H_2 + G_2G_3H_3) + G_1G_2G_3G_4H_1H_2$$

$$\Delta_1 = 1$$

$$T(s) = \frac{P_1\Delta_1}{\Delta}$$

$$\frac{C(s)}{R(s)} = \frac{G_1G_2G_3G_4}{1 + G_1G_2H_1 + G_3G_4H_2 - G_2G_3H_3 + G_1G_2G_3G_4H_1H_2}$$

Note that the overall transfer function that acquired from SFG is the same as the one that acquired from block diagram reduction.

Example 6.9 Obtain C(s)/R(s) for the SFG of Fig. 6.11 [2, p: 121, Q17].

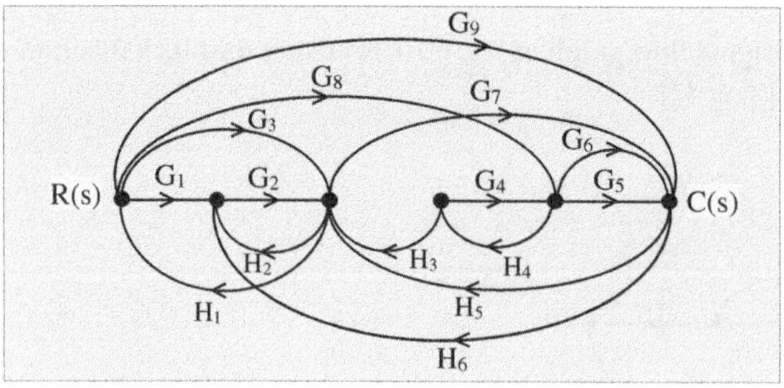

Figure 6.11 The signal flow graph of Example 6.9. Jairth [2] / Ane Books.

Solution:

The forward paths' gains:

$P_1 = G_1G_2G_7$ k: number of forward paths = 5

$P_2 = G_8G_5$

$P_3 = G_8G_6$

$P_4 = G_9$

$P_5 = G_3G_7$

Individual loops:

$L_1 = G_1G_2H_1$

$L_2 = G_2H_2$

$L_3 = G_7H_5$

$L_4 = G_4H_4$

$L_5 = G_2G_7H_6$

$L_6 = G_3H_1$

$L_7 = G_8H_4H_3H_1$

Two non-touching loops:

$L_1L_4 = G_1G_2G_4H_1H_4$

$L_2L_4 = G_2G_4H_2H_4$

$L_3L_4 = G_4H_4H_5G_7$

$L_5L_4 = G_2G_4H_4H_6G_7$

$L_6L_4 = G_3H_1G_4H_4$

Three non-touching loops:

Not exist in this example.

$$\Delta = 1 - (L_1 + L_2 + L_3 + L_4 + L_5 + L_6 + L_7) + (L_1L_4 + L_2L_4 + L_3L_4 + L_5L_4 + L_6L_4)$$

$$\Delta_1 = 1 - L_4 = 1 - G_4H_4$$

$$\Delta_2 = 1 - L_2 = 1 - G_2H_2$$

$$\Delta_3 = 1 - L_2 = 1 - G_2H_2$$

$$\Delta_4 = 1 - (L_2 + L_4) = 1 - (G_2H_2 + G_4H_4) = 1 - G_2H_2 - G_4H_4$$

$$\Delta_5 = 1 - L_4 = 1 - G_4H_4$$

$$T(s) = \frac{\sum_{k=1}^{5} P_k\Delta_k}{\Delta} = \frac{P_1\Delta_1 + P_2\Delta_2 + P_3\Delta_3 + P_4\Delta_4 + P_5\Delta_5}{\Delta}$$

Substitute for all expressions until getting T(S) in terms of G and H.

Example 6.10 Identify the following for the signal flow graph of Fig. 6.12.

- The forward paths and determine their gains.
- The individual loops and determine their gains.
- The self-loops and determine their gains.
- The two non-touching loops and determine their gains.
- The three non-touching loops and determine their gains.
- The four non-touching loops and determine their gains.

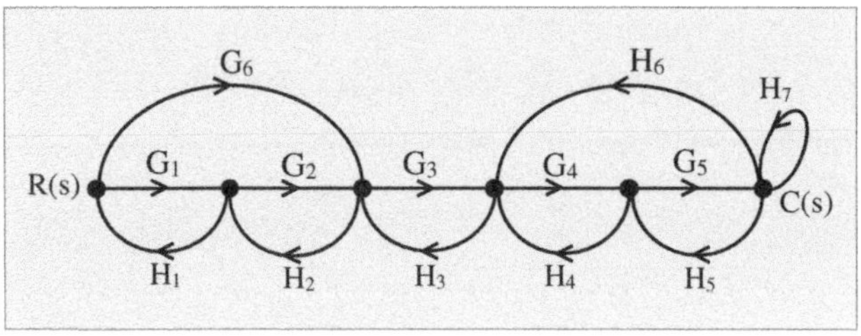

Figure 6.12 The signal flow graph of Example 6.10.

Solution:

The identifications of all the required paths and loops of the SFG of Fig. 6.12 are tabulated in the following table.

Forward paths	$P_1 = G_1G_2G_3G_4G_5$		
	$P_2 = G_6G_3G_4G_5$		
Individual loops	$L_1 = G_1H_1$	$L_4 = G_4H_4$	$L_7 = G_6H_2H_1$
	$L_2 = G_2H_2$	$L_5 = G_5H_5$	
	$L_3 = G_3H_3$	$L_6 = G_4G_5H_6$	
Self-loops	$L_8 = H_7$		
Two non-touching loops	$L_1L_3 = G_1G_3H_1H_3$	$L_2L_4 = G_2G_4H_2H_4$	$L_4L_7 = G_4H_4G_6H_2H_1$
	$L_1L_4 = G_1G_4H_1H_4$	$L_2L_5 = G_2G_5H_2H_5$	$L_4L_8 = G_4H_4H_7$
	$L_1L_5 = G_1G_5H_1H_5$	$L_2L_6 = G_2G_4G_5H_2H_6$	$L_5L_7 = G_5G_6H_2H_5H_1$
	$L_1L_6 = G_1G_4G_5H_1H_6$	$L_2L_8 = G_2H_2H_7$	$L_6L_7 = G_4G_5G_6H_2H_1H_6$
	$L_1L_8 = G_1H_1H_7$	$L_3L_5 = G_3G_5H_3H_5$	$L_8L_7 = G_6H_2H_1H_7$
		$L_3L_8 = G_3H_3H_7$	
Three non-touching loops	$L_1L_3L_5 = G_1G_3G_5H_1H_3H_5$		
	$L_1L_3L_8 = G_1G_3H_1H_3H_7$		
	$L_2L_4L_8 = G_2G_4H_2H_4H_7$		
	$L_1L_4L_8 = G_1G_4H_1H_4H_7$		
Four non-touching loops	Not exist in this example.		

Example 6.11 Identify the following for the signal flow graph of Fig. 6.13.

- Source node and sink node.
- The forward paths and determine their gains.
- The individual loops and determine their gains.
- The self-loops and determine their gains.
- The two non-touching loops and determine their gains.
- The three non-touching loops and determine their gains.
- The four non-touching loops and determine their gains.

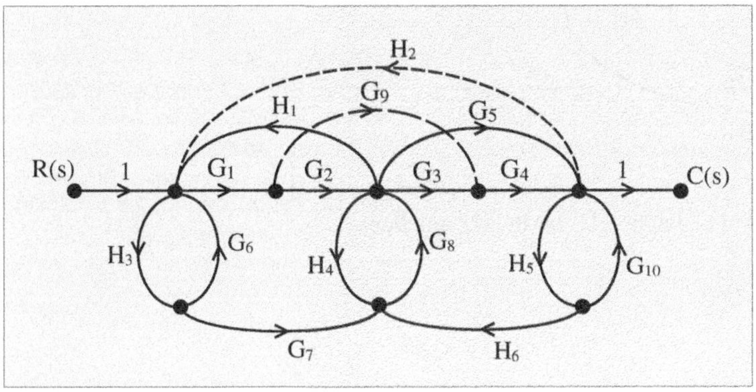

Figure 6.13 The signal flow graph of Example 6.11.

Solution:

The identifications for all the mentioned points about the SFG of Fig. 6.13 are illustrated in the following table.

Source node	The first node	
Sink node	The last node	
Forward paths	$P_1 = G_1G_2G_3G_4$	
	$P_2 = G_1G_9G_4$	
	$P_3 = G_1G_2G_5$	
Individual loops	$L_1 = G_1G_2H_1$	$L_5 = G_7G_8H_1H_3$
	$L_2 = G_6H_3$	$L_6 = G_1G_2G_3G_4H_2$
	$L_3 = G_8H_4$	$L_7 = G_3G_4H_5H_6G_8$
	$L_4 = G_{10}H_5$	
Self-loops	Not exist in this example.	
Two non-touching loops	$L_1L_4 = G_1G_2G_{10}H_1H_5$	$L_3L_4 = G_8G_{10}H_4H_5$
	$L_2L_3 = G_6G_8H_3H_4$	$L_4L_5 = G_7G_8G_{10}H_1H_3H_5$
	$L_2L_4 = G_6G_{10}H_3H_5$	
	$L_2L_7 = G_3G_4G_6G_8H_3H_5H_6$	
Three non-touching loops	$L_2L_3L_4 = G_6G_8G_{10}H_3H_4H_5$	
Four non-touching loops	Not exist in this example.	

Example 6.12 Obtain C(s)/R(s) for the signal flow graph of Fig. 6.14, knowing that the idea for this signal flow graph is taken from Jarith [2, p: 168, Fig. 7-11(c)].

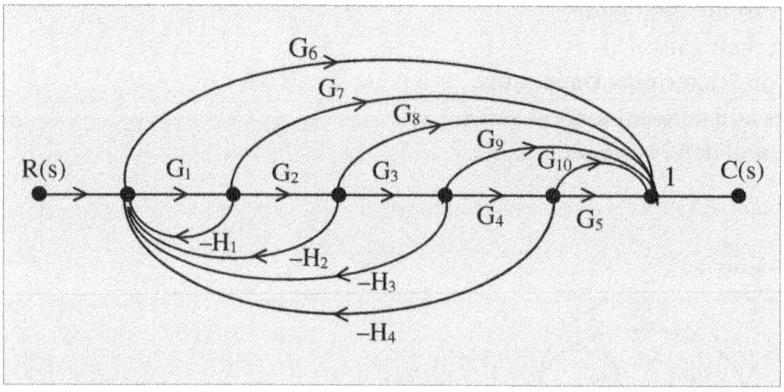

Figure 6.14 The signal flow graph of Example 6.12. Jairth [2] / Ane Books.

Solution:

The forward paths' gains:

$P_1 = G_1G_2G_3G_4G_5$ k: number of forward paths = 6

$P_2 = G_6$

$P_3 = G_1G_7$

$P_4 = G_1G_2G_8$

$P_5 = G_1G_2G_3G_9$

$P_6 = G_1G_2G_3G_4G_{10}$

Individual loops:

$L_1 = -G_1H_1$

$L_2 = -G_1G_2H_2$

$L_3 = -G_1G_2G_3H_3$

$L_4 = -G_1G_2G_3G_4H_4$

Two non-touching loops:

Not exist in this example.

Three non-touching loops:

Not exist in this example.

$$\Delta = 1 - (L_1 + L_2 + L_3 + L_4)$$

$$\Delta = 1 + G_1H_1 + G_1G_2H_2 + G_1G_2G_3H_3 + G_1G_2G_3G_4H_4$$

$$\Delta_1 = \Delta_2 = \Delta_3 = \Delta_4 = \Delta_5 = \Delta_6 = 1$$

$$T(s) = \frac{C(s)}{R(s)} = \frac{P_1\Delta_1 + P_2\Delta_2 + P_3\Delta_3 + P_4\Delta_4 + P_5\Delta_5 + P_6\Delta_6}{\Delta}$$

$$\boxed{\frac{C(s)}{R(s)} = \frac{G_1G_2G_3G_4G_5 + G_6 + G_1G_7 + G_1G_2G_8 + G_1G_2G_3G_9 + G_1G_2G_3G_4G_{10}}{1 + G_1H_1 + G_1G_2H_2 + G_1G_2G_3H_3 + G_1G_2G_3G_4H_4}}$$

Activity:

Obtain C(s)/U(s) for the signal flow graph of Fig. 9.7 on page 357 and Fig. 9.9 on page 358 [10].

Example 6.13 Obtain the overall transfer function for the SFG of Fig. 6.15.

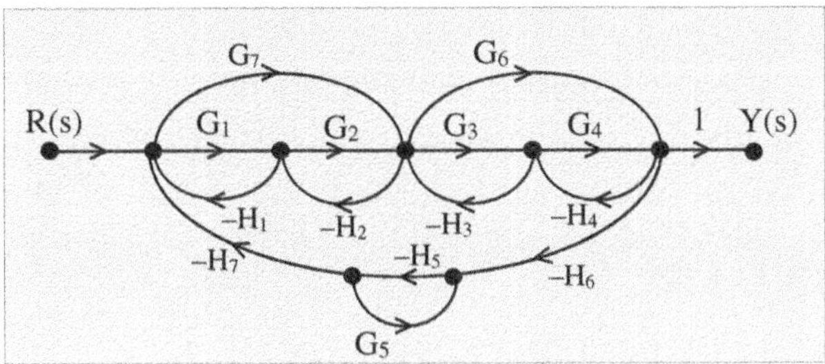

Figure 6.15 The signal flow graph of Example 6.13.

Solution:

The forward paths' gains:

$P_1 = G_1G_2G_3G_4$ k: number of forward paths = 4

$P_2 = G_6G_7$

$P_3 = G_1G_2G_6$

$P_4 = G_7G_3G_4$

Individual loops:

$L_1 = -G_1H_1$

$L_2 = -G_2H_2$

$L_3 = -G_3H_3$

$L_4 = -G_4H_4$

$L_5 = -G_5H_5$

$L_6 = -G_1G_2G_3G_4H_5H_6H_7$

$L_7 = G_7H_2H_1$

$L_8 = G_6H_4H_3$

Two non-touching loops:

$$L_1L_3 = G_1G_3H_1H_3 \qquad L_7L_5 = -G_7G_5H_2H_1H_5$$
$$L_1L_4 = G_1G_4H_1H_4 \qquad L_8L_5 = -G_6G_5H_4H_3H_5$$
$$L_1L_5 = G_1G_5H_1H_5$$
$$L_2L_4 = G_2G_4H_2H_4$$
$$L_2L_5 = G_2G_5H_2H_5$$
$$L_3L_5 = G_3G_5H_3H_5$$
$$L_4L_5 = G_4G_5H_4H_5$$

Three non-touching loops:

$$L_1L_3L_5 = -G_1G_3G_5H_1H_3H_5 \qquad L_7L_4L_5 = G_4G_5G_7H_1H_2H_4H_5$$
$$L_1L_4L_5 = -G_1G_4G_5H_1H_4H_5 \qquad L_8L_1L_5 = G_1G_5G_6H_1H_3H_4H_5$$
$$L_2L_4L_5 = -G_2G_4G_5H_2H_4H_5$$

Four non-touching loops:

Not exist in this example.

$$\Delta = 1 - (L_1 + L_2 + L_3 + L_4 + L_5 + L_6 + L_7 + L_8)$$
$$+ (L_1L_3 + L_1L_4 + L_1L_5 + L_2L_4 + L_2L_5 + L_3L_5 + L_4L_5 + L_7L_5 + L_8L_5)$$
$$- (L_1L_3L_5 + L_1L_4L_5 + L_2L_4L_5 + L_7L_4L_5 + L_8L_1L_5)$$

$$\Delta_1 = 1 - L_5 = 1 + G_5H_5, \quad \Delta_2 = 1 - L_5 = 1 + G_5H_5 = \Delta_3 = \Delta_4$$

$$T(s) = \frac{C(s)}{R(s)} = \frac{\sum_{k=1}^{4} P_k\Delta_k}{\Delta} = \frac{P_1\Delta_1 + P_2\Delta_2 + P_3\Delta_3 + P_4\Delta_4}{\Delta}$$

Simplify until getting the overall transfer function.

For More Practice:

- See Chapter 7 on pages 177–214 [4]. There are many solved examples that have separate graphs for each forward path to confirm understanding.
- See example 10 on page 45 [1].

6.7 Converting A Block Diagram Into A Signal Flow Graph

Any block diagram can be easily converted into an SFG. Each summing point or takeoff point in the block diagram can be represented by a node in the SFG. All G's and H's blocks can be represented by drawing lines to connect these nodes. Finally, draw arrows on those lines and place plus or minus signs to represent the signal flow if it is positive or negative. Table 6.2 illustrates this process.

Table 6.2 Block diagram and its equivalent signal flow graph illustrations.

Block diagram	**Equivalent signal flow graph**
Summing point or takeoff point 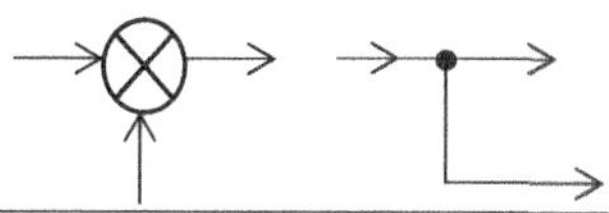	Node 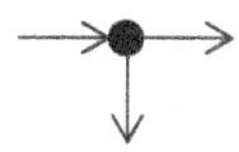
Block 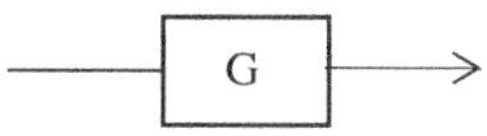	Line 
Summing point & takeoff point as separated nodes 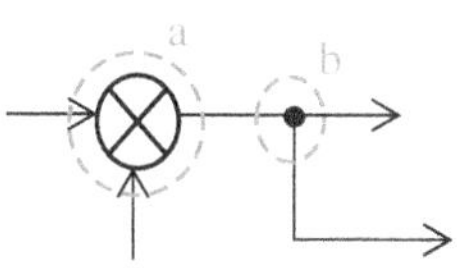	Two nodes 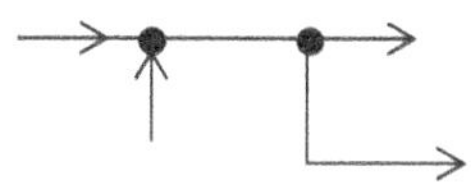
Summing point & takeoff point as one node 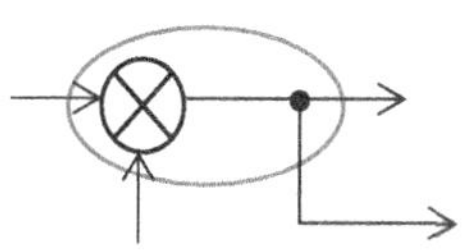	One node 
Takeoff point & summing point as separated nodes 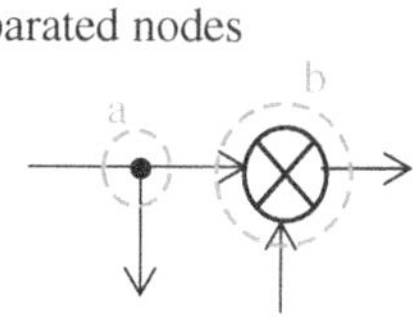	Two nodes 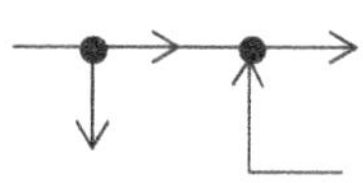
Takeoff point & summing point as one node 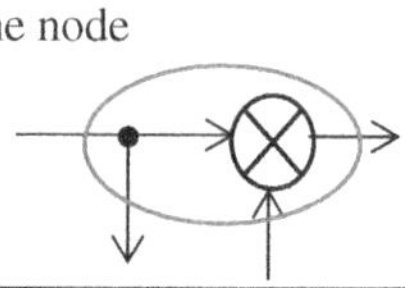	One node 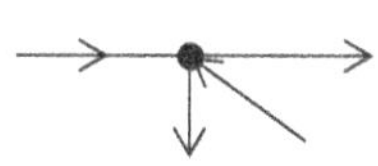

6.8 Examples of Converting a Block Diagram Into Its Equivalent SFG

Example 6.14 Convert the block diagram of Fig. 6.16 to its equivalent signal flow graph [3, p: 60, Fig. 2-31].

Solution:

The block diagram of Fig. 6.16 consists of five variables, as demonstrated by a, b, c, d, and e, which show how the signal is transferred from the input to the output. These variables can be represented

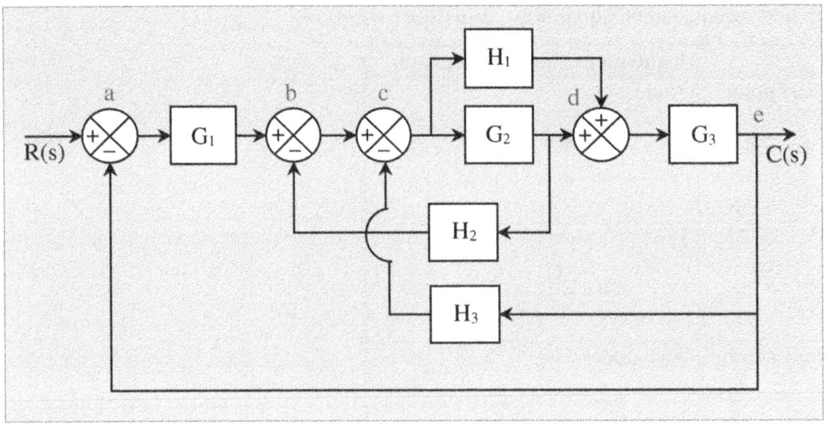

Figure 6.16 The block diagram of Example 6.14. Adapted from Ogata [3].

by nodes (a, b, c, d, and e) in SFG starting from the source node and ending with the output node. The lines and arrows in the SFG show the forward paths or feedback paths that connect the blocks in the block diagram. So, the final SFG for the block diagram of Fig. 6.16 will be as Fig. 6.16-a.

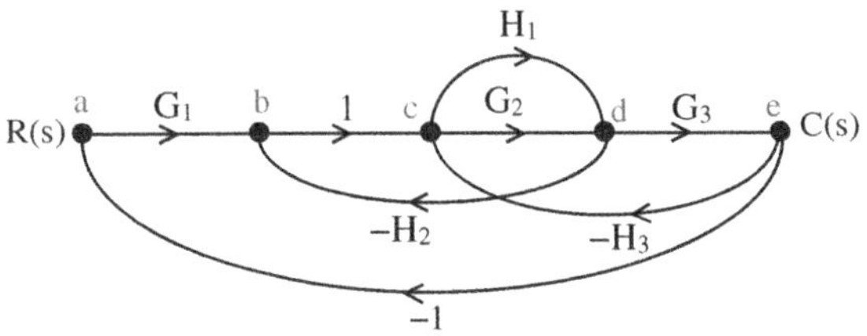

Figure 6.16-a The signal flow graph for the block diagram of Example 6.14.

Example 6.15 Convert the block diagram of Fig. 6.17 into its equivalent signal flow graph [8, p: 280, Fig. P5-6].

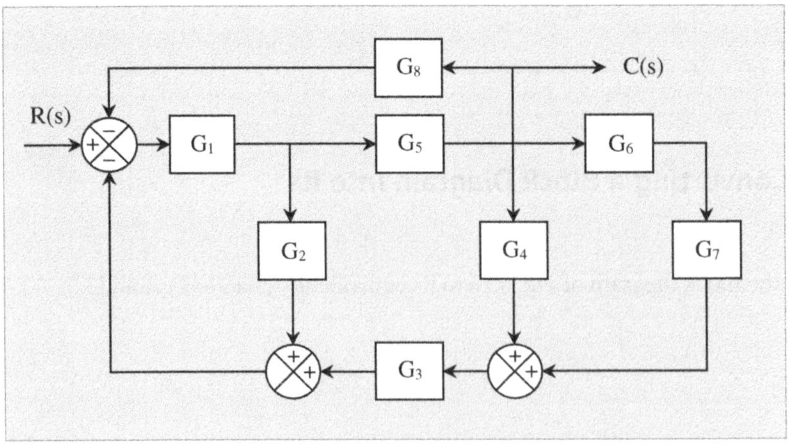

Figure 6.17 The block diagram of Example 6.15. Adapted from Nise [8].

Solution:

By following the same process as illustrated in Example 6.14, the equivalent SFG of the block diagram of Fig. 6.17 should look like Fig. 6.17-a.

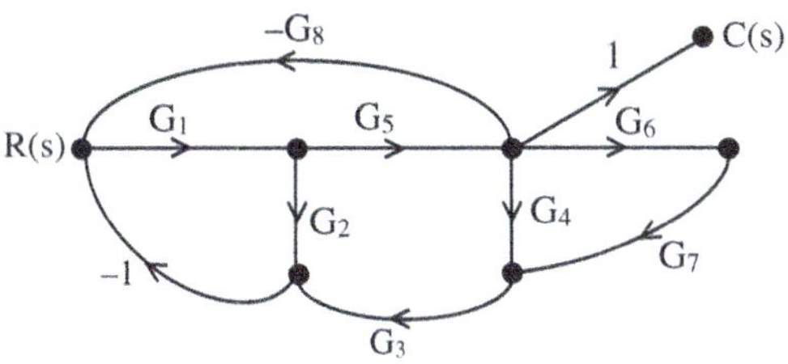

Figure 6.17-a The signal flow graph for the block diagram of Example 6.15.

Example 6.16 Convert the block diagram of Fig. 6.18 into its equivalent signal flow graph.

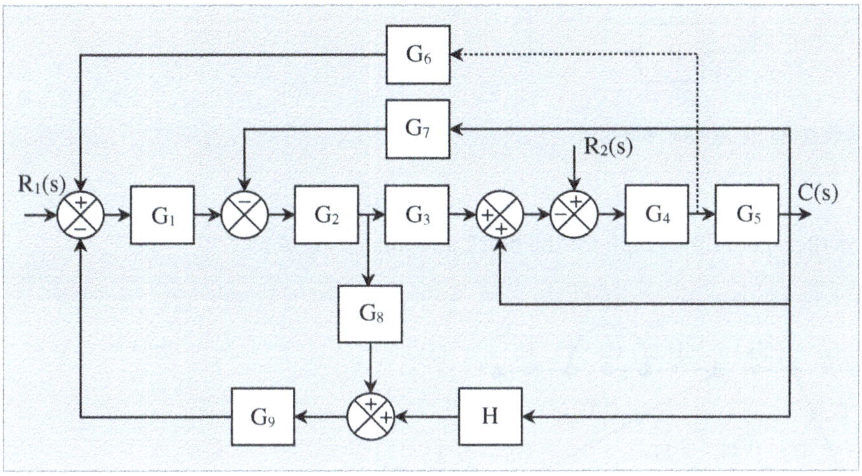

Figure 6.18 The block diagram of Example 6.16.

Solution:

The SFG of the block diagram of Fig. 6.18 is shown in Fig. 6.18-a. Note that we add 3 nodes to make the source nodes and the output node clear, and place the gain 1 on the lines.

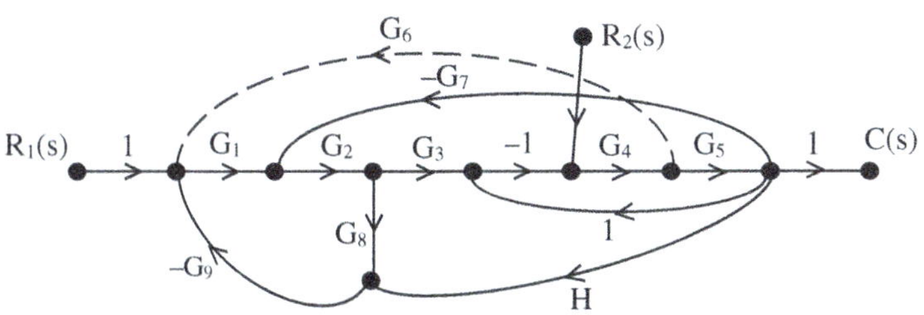

Figure 6.18-a The signal flow graph for the block diagram of Example 6.16.

Example 6.17 Convert the two-input/two-output system block diagram of Fig. 6.19 into its equivalent signal flow graph, then obtain:

$$\frac{Y_1(s)}{X_1(s)}, \frac{Y_2(s)}{X_1(s)}, \frac{Y_1(s)}{X_2(s)}, \frac{Y_2(s)}{X_2(s)} \text{ transfer functions.}$$

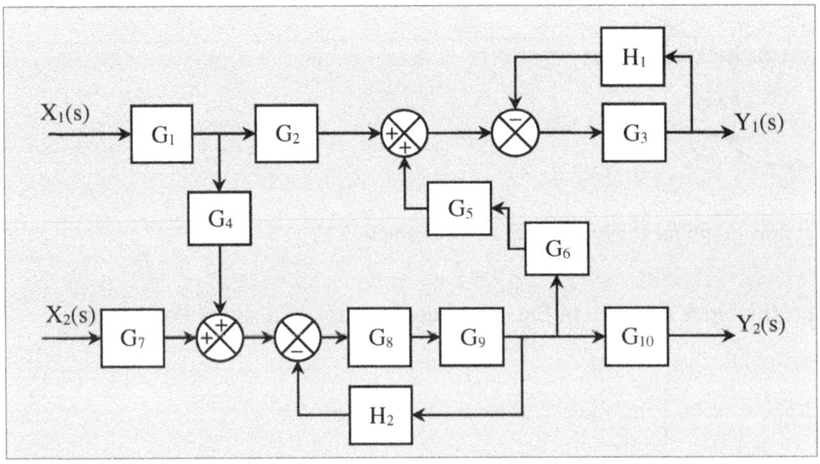

Figure 6.19 The block diagram of Example 6.17.

Solution:

1- Fig. 6.19-a shows the signal flow graph for the block diagram of Fig. 6.19.

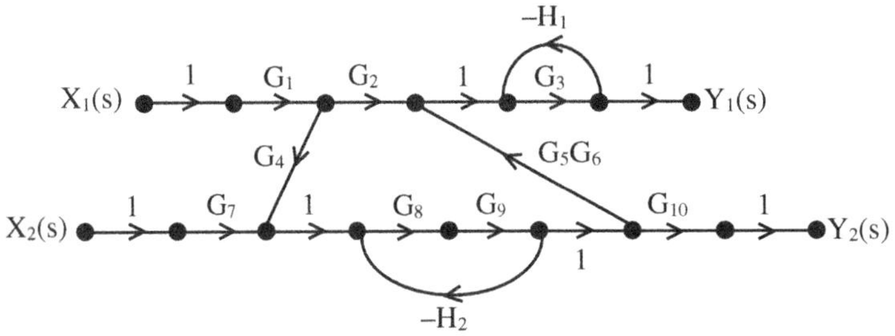

Figure 6.19-a The signal flow graph for the block diagram of Fig. 6.19.

2- To get $Y_1(s)/X_1(s)$: the signal flow graph of Fig. 6.19-a will be as shown in Fig. 6.19-b.

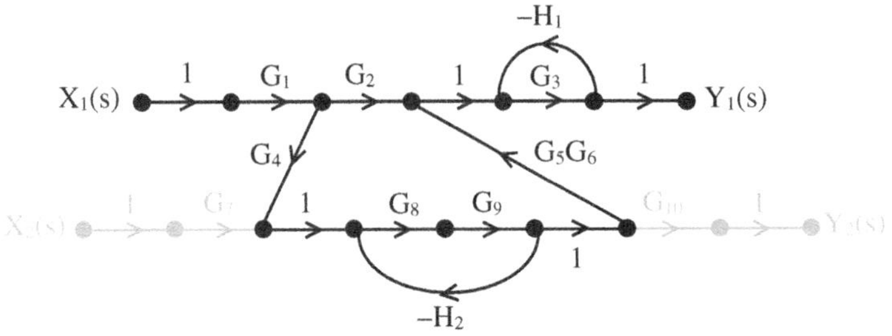

Figure 6.19-b The signal flow graph of step 2.

The forward paths' gains:

$P_1 = G_1G_2G_3$ k: number of forward paths = 2

$P_2 = G_1G_4G_8G_9G_5G_6G_3$

Individual loops:

$L_1 = -G_3H_1$

$L_2 = -G_8G_9H_2$

Two non-touching loops:

Not exist in this example.

Three non-touching loops:

Not exist in this example.

$$\Delta = 1 - (L_1 + L_2)$$

$$\Delta = 1 - (-G_3H_1 - G_8G_9H_2) = 1 + G_3H_1 + G_8G_9H_2$$

$$\Delta_1 = 1 - L_2 = 1 + G_8G_9H_2$$

$$\Delta_2 = 1$$

$$P_1\Delta_1 = G_1G_2G_3(1 + G_8G_9H_2)$$

$$P_2\Delta_2 = G_1G_4G_8G_9G_5G_6G_3$$

$$\frac{Y_1(s)}{X_1(s)} = \frac{\sum_{k=1}^{2} P_k\Delta_k}{\Delta} = \frac{P_1\Delta_1 + P_2\Delta_2}{\Delta}$$

$$\frac{Y_1(s)}{X_1(s)} = \frac{G_1G_2G_3(1 + G_8G_9H_2) + G_1G_4G_8G_9G_5G_6G_3}{1 + G_3H_1 + G_8G_9H_2}$$

3- o get $Y_2(s)/X_1(s)$: the signal flow graph of Fig. 6.19-a will be as shown in Fig. 6.19-c.

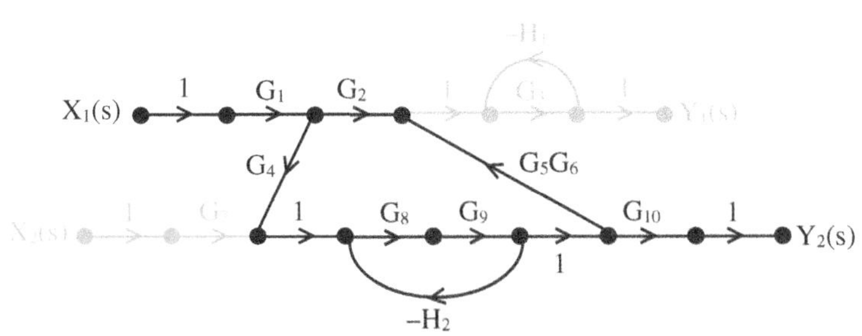

Figure 6.19-c The signal flow of step 3.

The forward path gain:

$P_1 = G_1G_4G_8G_9G_{10}$ k: number of forward paths = 1

Individual loops:

$$L_1 = -G_8G_9H_2$$

Two non-touching loops:

Not exist in this example.

Three non-touching loops:

Not exist in this example.

$$\Delta = 1 - L_1 = 1 + G_8G_9H_2$$

$$\Delta_1 = 1$$

$$P_1\Delta_1 = G_1G_4G_8G_9G_{10}$$

$$\frac{Y_2(s)}{X_1(s)} = \frac{P_1\Delta_1}{\Delta}$$

$$\boxed{\frac{Y_2(s)}{X_1(s)} = \frac{G_1G_4G_8G_9G_{10}}{1 + G_8G_9H_2}}$$

4- To get $Y_1(s)/X_2(s)$: the signal flow graph of Fig. 6.19-a will be as shown in Fig. 6.19-d.

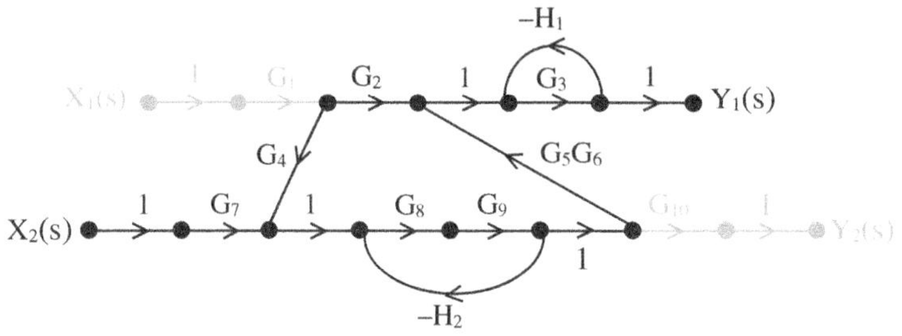

Figure 6.19-d The signal flow graph of step 4.

The forward path gain:

$P_1 = G_7G_8G_9G_6G_5G_3$ k: number of forward paths = 1

Individual loops:

$$L_1 = -G_3H_1$$
$$L_2 = -G_8G_9H_2$$

Activity:

- Continue the solution until you get $Y_1(s)/X_2(s)$ transfer function.
- Draw the signal flow graph for the last case and follow the same process to get $Y_2(s)/X_2(s)$ transfer function.

Example 6.18 Convert the block diagram of Fig. 6.20 into signal flow graph, then obtain the overall transfer function.

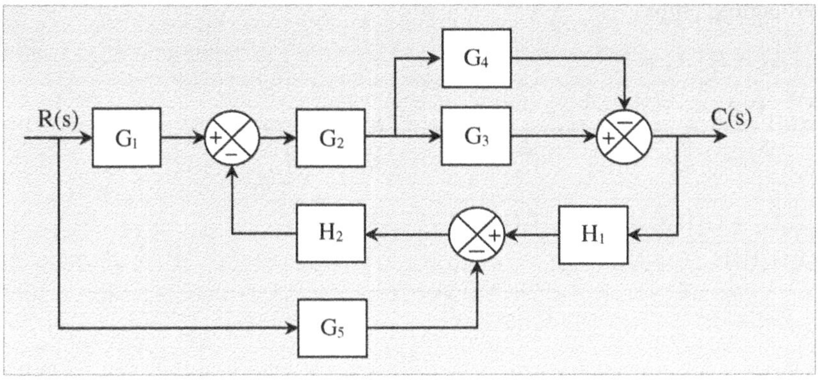

Figure 6.20 The block diagram of Example 6.18.

Solution:

The SFG for the block diagram of Fig. 6.20 will be as in Fig. 6.20-a

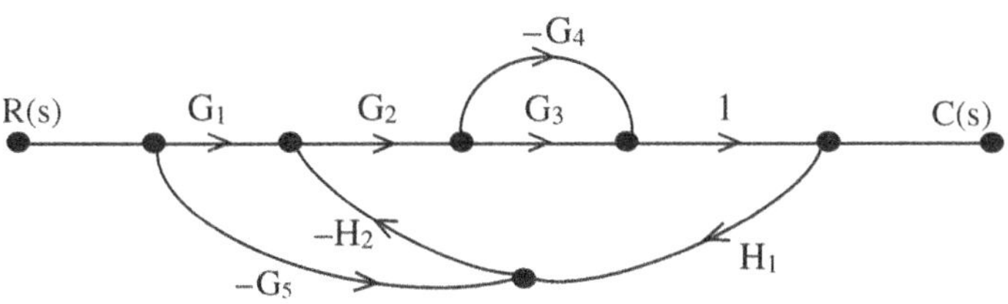

Figure 6.20-a The signal flow graph for the block diagram of Fig. 6.20.

The forward paths' gains:

$P_1 = G_1G_2G_3$ k: number of forward paths = 4

$P_2 = -G_1G_2G_4$

$P_3 = G_5H_2G_2G_3$

$P_4 = -G_5H_2G_2G_4$

Individual loops:

$L_1 = -G_2G_3H_1H_2$

$L_2 = G_2G_4H_1H_2$

Two non-touching loops:

Not exist in this example.

Three non-touching loops:

Not exist in this example.

$$\Delta = 1 - (L_1 + L_2)$$

$$\Delta = 1 + G_2G_3H_1H_2 - G_2G_4H_1H_2$$

Because there are no touching loops.

$$\Delta_1 = 1,\ \ \Delta_2 = 1,\ \ \Delta_3 = 1,\ \ \Delta_4 = 1$$

$$T(s) = \frac{C(s)}{R(s)} = \frac{\sum_{k=1}^{4} P_k\Delta_k}{\Delta} = \frac{P_1\Delta_1 + P_2\Delta_2 + P_3\Delta_3 + P_4\Delta_4}{\Delta}$$

$$\boxed{\frac{C(s)}{R(s)} = \frac{G_1G_2G_3 - G_1G_2G_4 + G_5H_2G_2G_3 - G_5H_2G_2G_4}{1 + G_2G_3H_1H_2 - G_2G_4H_1H_2}}$$

Example 6.19 Convert the two-input/single-output system of Fig. 6.21 into its equivalent SFG and obtain $\frac{C(s)}{R(s)}$ and $\frac{C(s)}{D(s)}$ transfer functions [3, P: 61, Fig. 2 34].

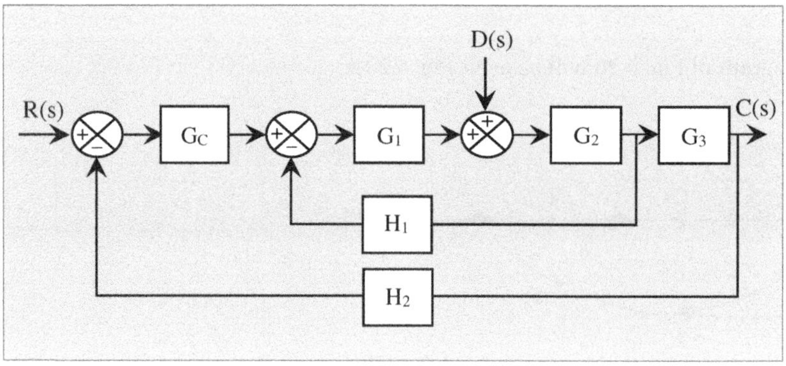

Figure 6.21 The block diagram of Example 6.19. Adapted from Ogata [3].

Solution:

The SFG for the block diagram of Fig. 6.21 will be as Fig. 6.21-a.

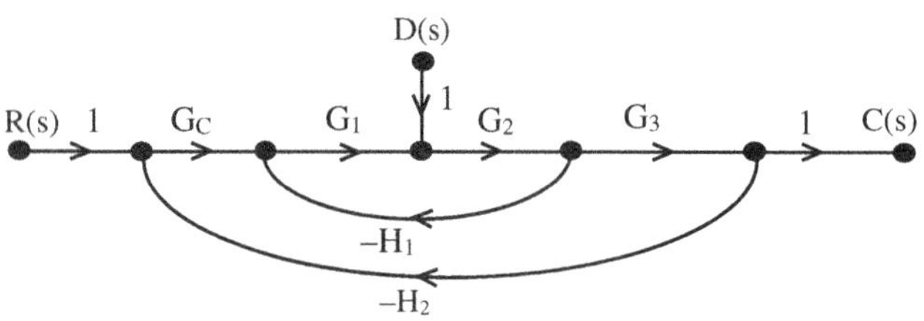

Figure 6.21-a The signal flow graph for the block diagram of Fig. 6.21.

The transfer function can be easily found by applying the superposition theorem, one input at a time. So, when $D(s) = 0$ and $R(s)$ exist, Fig. 6.21-a will be as Fig. 6.21-b.

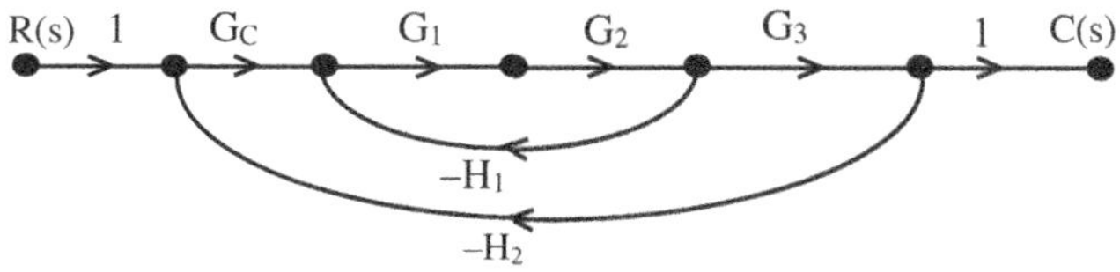

Figure 6.21-b The signal flow graph of Fig. 6.21-a when D(s) = 0.

The forward path gain:

$P_1 = G_C G_1 G_2 G_3$ k: number of forward paths = 1

Individual loops:

$L_1 = -G_1 G_2 H_1$

$L_2 = -G_C G_1 G_2 G_3 H_2$

$\Delta = 1 - (L_1 + L_2) = 1 + G_1 G_2 H_1 + G_C G_1 G_2 G_3 H_2$

$\Delta_1 = 1$

$$\frac{C(s)}{R(s)} = \frac{P_1 \Delta_1}{\Delta} = \frac{G_C G_1 G_2 G_3}{1 + G_1 G_2 H_1 + G_C G_1 G_2 G_3 H_2}$$

when R(s) = 0 and D(s) exist, Fig. 6.21-a will be as Fig. 6.21-c.

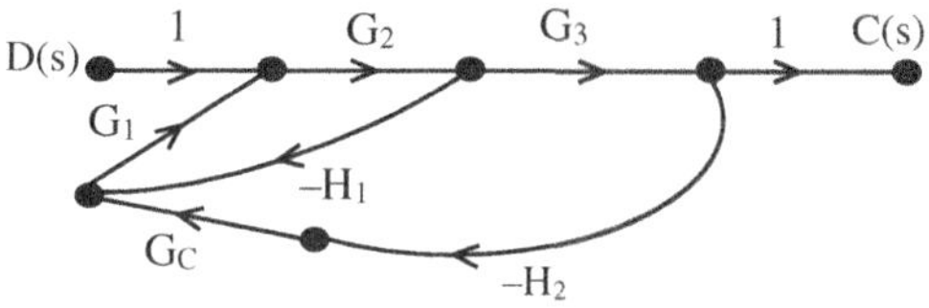

Figure 6.21-c The signal flow graph of Fig. 6.21-a when R(s) = 0.

The forward path gain:

$P_1 = G_2 G_3$ k: number of forward paths = 1

Individual loops:

$L_1 = -G_1 G_2 H_1$

$L_2 = -G_2 G_3 H_2 G_C G_1$

$\Delta = 1 - (L_1 + L_2) = 1 + G_1 G_2 H_1 + G_2 G_3 H_2 G_C G_1$

$\Delta_1 = 1$

$$\frac{C(s)}{D(s)} = \frac{P_1 \Delta_1}{\Delta} = \frac{G_2 G_3}{1 + G_1 G_2 H_1 + G_2 G_3 H_2 G_C G_1}$$

Example 6.20 Convert the block diagram of Fig. 6.22 into its equivalent SFG. [3, P: 48, Fig. 2-21 a].

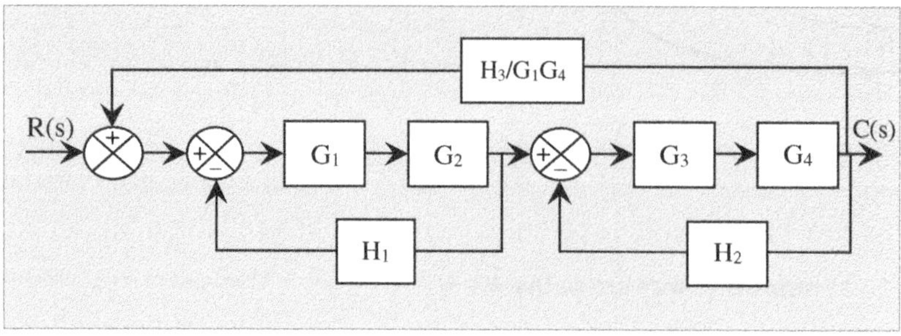

Figure 6.22 The block diagram of Example 6.20. Adapted from Ogata [3].

Solution:

The SFG for the block diagram or Fig. 6.22 will be as shown in Fig. 6.22-a.

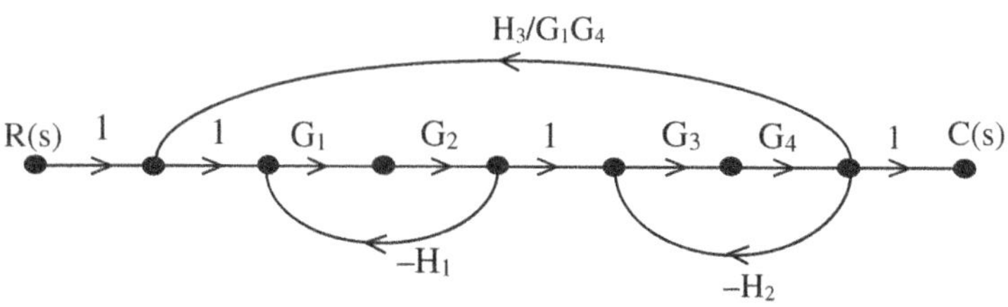

Figure 6.22-a The signal flow graph for the block diagram of Fig. 6.22.

For More Practice:

Go to references, read and solve all the examples and problems.

Review Questions

1 Obtain the overall transfer function for the signal flow graphs of Fig. 6.23 [9, P: 142, Fig. 3P-26 a].

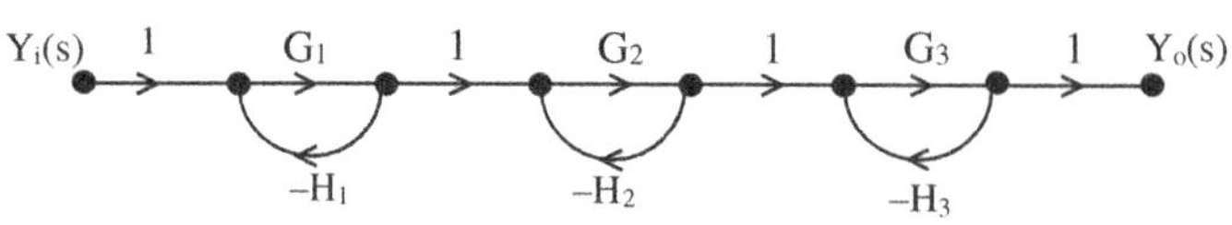

Figure 6.23 The signal flow graph of Q.1.

2 Obtain the overall transfer function for the signal flow graphs of Fig. 6.24, then compare the results with the previous question [9, P: 142, Fig. 3P-26 b].

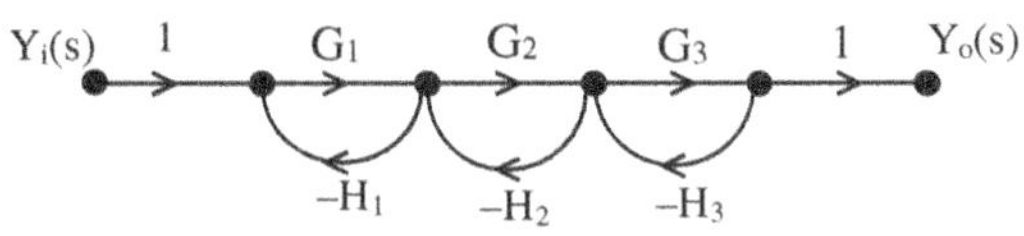

Figure 6.24 The signal flow graph of Q.2.

3 Find the transfer function of the system shown in Fig. 6.25 by using Mason's gain formula. [9, p: 140, Fig. 3p-22 a].

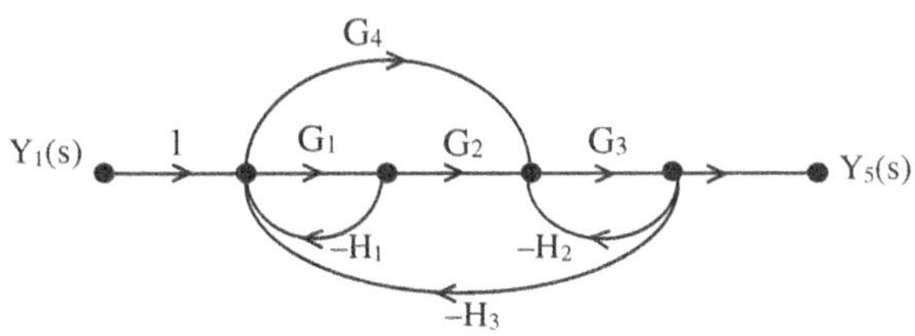

Figure 6.25 The signal flow graph of Q.3.

4 Find the transfer function of the system shown in Fig. 6.26 by using Mason's gain formula. [9, p: 140, Fig. 3p-22 c].

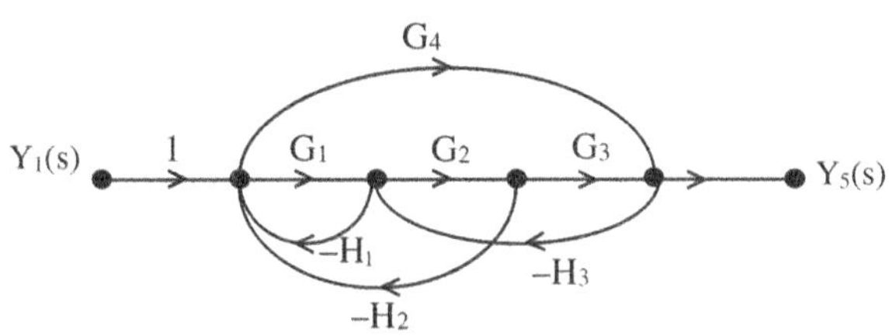

Figure 6.26 The signal flow graph of Q.4.

5 Convert the block diagram of Fig. 6.27 into its equivalent SFG. [9, P: 115, Fig. 3.18-b].

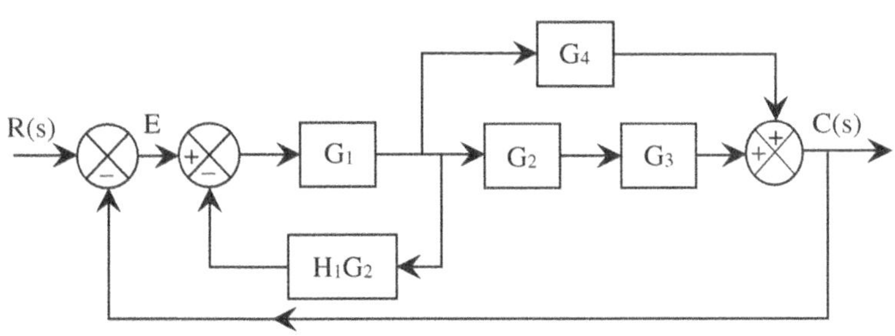

Figure 6.27 The block diagram of Q.5.

Appendix

Loops:

A loop is a method for repeating a computer program several times. MATLAB has different types of loops, one of them is `for-end` loop [6, p: 208].

The `for-end` Structure:

Fig. A.1 shows the structure of `for-end` loop. `For` has a counter i followed by a group of MATLAB commands or statements. The counter i has start and end values that show how many times the program will be executed.

```
for i=f:s:L
      -----  \
      -----   )  A Group of MATLAB
     ------  /   commands or statements
 end
```

Figure A.1 `for-end` loop structure.

Where

`i` `is` the counter index
`f` is the first value for the counter i
`s` is the step. If you don't set a number for s, MATLAB will consider it 1 as a default number.
`L` is the last value of the counter i.

Nested loop [6, p: 216]

Conditional statements and loops in MATLAB can be nested with themselves and one another as illustrated in Fig. A.2.

Block Diagram Implementation for Control Systems: Including MATLAB® Programming, Microsoft® Word and PowerPoint Presentation, First Edition. Ruba Al-Mulla Hummadi.

```
for i=1:n                         ┐
    for j=1:m      ┐              │ Loop
        -----      │ Nested       │
        -----      │ loop         │
       ------      │              │
    end            ┘              │
end                               ┘
```

Figure A.2 Nested conditional statements and loops.

Conditional Statement [6, p: 200]

A conditional statement is a command that permits MATLAB to decide whether to execute a sequence of commands that follow the conditional statement or leave them. It could be in different structures depending on the condition(s) like:

Form	Usage
`If-end`	for one condition
`If-else-end`	for two conditions
`If-elseif-else-end`	more than two conditions

`If-elseif-else-end` Structure:

`If` condition expression
  Statement(s)
`elseif` another conditional expression
  Statement(s)
`else` no more conditions
  Statement(s)
`end`

Now, let us return to Example 2.1 in Chapter 2.

```
clear
clc
% % file name: math.m
A =[1 2 4;3 5 1; -4 2 -3];     input data
B =[2 5 7;8 -1 3; -5 4 2];
c = A+B
c1=4*A
c2=B(1,:)
for i=1:3                      ← loop
    for j=1:3                  ← nested loop
        if i==j                ← if condition
            c3(1,j)= A(i,j);   ← statement
        end                    ← end of condition
    end
end
c3
c4= A*B+c1
```

References

1 S. Panda, S. Padhy, and S. R. Das, *Control System Engineering*. New Delhi: S.K. Kataria & Sons, 2013.

2 A. K. Jairth, *Control Systems – The State Variable Approach (Conventional and MATLAB)*. New Delhi: Ane Books Pvt. Ltd, 2010.

3 K. Ogata, *Modern Control Engineering*, 5th ed. New Jersey: Prentice Hall, 2010.

4 S. Ghosh, *Control Systems Theory and Applications*. New Delhi: Dorling Kindersley (India) Pvt. Ltd, 2007.

5 W. J. Palm III, *Introduction to MATLAB for Engineers*, 3rd ed. New York: McGraw Hill, 2011.

6 A. Gilat, *MATLAB An Introduction with Applications*, 3rd ed. United States of America: John Wiley & Sons, Inc, 2008.

7 F. H. Raven, *Automatic Control Engineering*, 3rd ed. McGraw-Hill Inc., 1961.

8 N. S. Nise, *Control System Engineering*, 7th ed. New Jersey: John Wiley & Sons, 2015.

9 B. C. Kuo and F. Golnaraghi, *Automatic Control Systems*, 9th ed. USA: John Wiley & Sons Inc, 2010.

10 R. C. Dorf, *Modern Control Systems*, 4th ed. Reading, Massachusetts: Addison-Wesley Publishing Company, 1986.

Additional References

1 Go online and search for books that deal with Control Systems then read the examples. Some of these books are free downloaded.

2 Go on YouTube and search for block diagram reduction and SFG examples. Teachers in the videos explain the whole question and you can pause the video whenever you feel you want to.

3 Go to the library and ask for Control books and MATLAB books then read their examples.

4 Go on YouTube and search for anything you want about MATLAB then watch the video.

5 Use MATLAB help. It is very helpful. Just write anything you want to be helped with after the word help in the Command Window then press enter. For example, help series ↵. MATLAB will show and explain how to use the series command.

Block Diagram Implementation for Control Systems: Including MATLAB® Programming, Microsoft® Word and PowerPoint Presentation, First Edition. Ruba Al-Mulla Hummadi.

Index

Block Diagram Implementation for Control Systems: Including MATLAB® Programming, Microsoft® Word and PowerPoint Presentation, First Edition. Ruba Al-Mulla Hummadi.